The Analyst

The Analyst

A Daughter's Memoir

Alice Wexler

Columbia University Press

New York

Columbia University Press
Publishers Since 1893
New York   Chichester, West Sussex
cup.columbia.edu

Library of Congress Cataloging-in-Publication Data
Names: Wexler, Alice, 1942– author.
Title: The analyst: a daughter's memoir / Alice Wexler.
Description: New York: Columbia University Press, [2022] | Includes
bibliographical references and index.
Identifiers: LCCN 2022006276 (print) | LCCN 2022006277 (ebook) | ISBN
9780231202787 (hardback) | ISBN 9780231554718 (ebook)
Subjects: LCSH: Wexler, Milton, 1908–2007. | Wexler, Alice, 1942– |
Psychoanalysts—United States—Biography. | Psychoanalysts—Family
relationships—United States—Biography.
Classification: LCC RC438.6.W49 W49 2022 (print) | LCC RC438.6.W49
(ebook) | DDC 616.89/170092 [B]—dc23/eng/20220223
LC record available at https://lccn.loc.gov/2022006276
LC ebook record available at https://lccn.loc.gov/2022006277

Cover design: Milenda Nan Ok Lee
Cover photo: Nancy S. Wexler

But even with respect to the most insignificant things in life, none of us con-stitutes a material whole, identical for everyone, which a person has only to go look up as though we were a book of specifications or a last testament; our social personality is a creation of the minds of others.

—Marcel Proust, *Swann's Way*

Contents

The Analyst

Prologue

I n the depths of the Great Depression, when many New York Jews of his generation were drawn to Marx, my father, a young lawyer, was drawn to Freud. He left the law to become a psychoanalyst at a time when, for many people, psychoanalysis was a scandal, associated with sex and unspeakable desires. Freud's idea that we are hostage to unconscious longings, accessible only through dreams and free associations, offended bourgeois notions of self-control. His terms for a divided mind, at least when translated into English, seemed alien, even though they harked back to ancient wisdom from poetry and myth, for instance when he spoke of sex and aggression emanating from the id; the puny kingdom of reason he labeled the ego; and an often brutal arena of conscience and morality he named the superego. But however strange these and other Freudian terms appeared to many people, the new profession based on them soon acquired great cachet, claiming to offer not only a theory of the mind but also a path to self-knowledge and a method of healing mental ills. Psychoanalysts enjoyed enormous cultural influence from the 1940s to the early 1960s—the so-called golden age of psychoanalysis. Growing up during those years, my sister, Nancy, and I thought psychoanalysis was glamorous and exciting even if psychoanalysts, with a few exceptions, were not. We thought our father was one of those exceptions, and quite a few others did as well.

Freud remained a touchstone for my father and psychoanalysis the bedrock of his identity. Yet much of his practice lay outside the borders of mainstream analysis, from the experimental therapy of schizophrenia he developed at the famed Menninger Foundation in the late forties and early fifties to his therapy groups with artists in the sixties to the freewheeling interdisciplinary workshops he initiated with scientists in the seventies. Although not quite a rebel, Dad was always a maverick, a PhD among the MDs; an orthodox Freudian who rejected notions of conformity and adjustment to the status quo, who believed in dreaming big and acting bold yet rarely lost sight of the practical world; a charismatic figure who, late in his life, crossed ethical boundaries in friendships and fundraising with patients; a husband and father and ex-husband, too, trying to weigh the balance between his responsibilities to others and his desire to live life on his own terms.

After my mother was diagnosed with what was then called Huntington's chorea, a hereditary malady of movement, mind, and mood, and my father knew my sister and I each had a 50 percent chance of inheriting her deadly disease and of passing it on to any children we might have, he changed his life again. He dedicated himself to advocacy for biomedical research even as he continued to practice therapy. He began to think of himself less as a therapist and more as a teacher or philosopher or even fellow sufferer. He told his patients that he was prepared to struggle with them in an effort to improve the shape of their lives. He continued to question himself and his choices, to tell and retell his story. Yet he never quite found a narrative that did justice to the complexity, richness, and contradictions of his ninety-eight years.

As children do, I have questions about my father that have never been answered. I wanted to learn more of his wisdom and understand how he saw his life, looking at the world and himself as he did, through a Freudian lens. I write this book not only as a daughter but also as a historian and biographer, to situate my father in the larger stories of psychoanalysis and self-making, of families and fatherhood, in twentieth-century America. And I write to explore the circumstances that enabled him to go beyond his situation—as the philosopher of existentialism Jean-Paul Sartre put it, to make something of what he had been made.

In a previous book, *Mapping Fate*, I wrote our family story with a focus on my mother, in accord with Virginia Woolf's feminist maxim, "We think back through our mothers if we are women." In writing this book I have come to appreciate that I am my father's daughter as well.

I

On the Road to Topeka

*If the Archive is a place of dreams, it permits this one, above all others,
the one that Michelet dreamed first, of making the dead walk and talk.*

—Carolyn Steedman, *Dust*

I'm driving west along Interstate 70 from the Kansas City airport on a mild fall afternoon in 2014 when the weather report comes over the radio: a polar vortex is blowing in, bringing frigid Arctic air to the Midwest. Temperatures are about to sink below freezing. By the time I reach the Hyatt Place hotel in Topeka, the thermometer in my rental car registers fourteen degrees Fahrenheit. I feel anxious about driving on icy city streets, something I have not done for forty years.

But I am excited, too. I've come to Topeka to look once again at the place that made a deep impact on my father and shaped the childhood of my sister and me, this state capital where we lived in the years just before the landmark 1954 Supreme Court case *Brown v. Board of Education of Topeka* struck down segregation in America's public schools. We heard not a word nor did we think about segregation in Miss Snyder's all-white third grade at Randolph Elementary in 1951. Nor did Miss Snyder tell us about Topeka's role as a storied stop on the Underground Railroad or about the Exodusters, those African Americans from former slave states who fled the violence of the post-Reconstruction South to seek new lives in Kansas, a state idealized at the time for its antislavery past. That historical Kansas seems far away even as John Steuart Curry's magnificent mural in the Kansas statehouse explodes in my face with its towering figure of the abolitionist John Brown, arms outstretched, before a looming tornado funnel symbolizing the approach of the Civil War. I wonder if my father saw this mural when he first came to Topeka and what he may have thought of it.

Mostly I've come to immerse myself in the archives of that unique assemblage of institutions known as Menninger, which Charles Menninger, a family physician, and his forward-looking psychiatrist son, Karl, founded here in the heart of the Midwest, joined within a few years by another son, William (Will). The Menninger Clinic came first, in 1919, followed by a private psychiatric hospital, the Southard School for children, the Menninger Foundation, and Schools of Psychiatry, Clinical Psychology, and Nursing, all affiliated with the Winter Veterans Administration (VA) Hospital, which in 1946 was turned from a general army hospital into the VA's pilot hospital for neuropsychiatric training. Karl and Will Menninger, Kansas born and bred, hoped to do for psychiatry and psychoanalysis what the Mayo brothers had done for general medicine: to found a major center for treatment and especially for training and research. They wanted to bring into their programs the spirit of Kansas's progressive reform traditions and the idea of "we" instead of "me," as one graduate put it. And in the prosperous post–World War II years of generous funding for mental health treatment and research, inspired by the massive needs of traumatized veterans returning from battlefields around the world and by the growing prestige of science generally, the Menninger brothers turned Topeka into a legendary destination, one that helped shape mental health care in America for decades to come.[1]

Menninger is now but a memory here, the foundation and hospital closed, the clinic moved to Houston, Texas, where it is affiliated with the Baylor College of Medicine. But the historic Menninger Clinic building, purchased in 1925, still stands on West Sixth Street. The abandoned West Campus sits high on a hill overlooking the city, the old Menninger Hospital clocktower building stationed like a sentinel guarding invisible subjects. To the south, on West Twentieth Street, our former home looks much as I remember it, a compact, two-story, white, wood-frame house—although the unfamiliar picket fence out front now accentuates its prim Dick-and-Jane appearance. The mulberry tree that my sister and I loved to climb still stands in the wintry front yard, and magnolia trees laden with large pink blossoms line this street without sidewalks, glorious holdouts against the bitter cold.

Even before I returned to Topeka that afternoon I had felt inchoately that my father's five years at Menninger were the key to his entire life. He always spoke of Menninger (we always called it "Menninger's," the local nomenclature) with a kind of reverence as a place of intellectual brilliance and emotional intensity,

FIGURE 1.1 The old Menninger Clinic building, acquired in 1925 and still standing on West Sixth Street, 2014.

where he worked with colleagues he admired and respected and where he developed a sense of self-confidence and command, as he put it, and a feeling that there wasn't much that other people were doing that he could not do himself. I was too young during those years to remember much of Topeka. Most of what I knew about my father's professional life at Menninger came from the stories he told us or wrote down many years after we left, stories created anew in the way that stories are each time we retrieve them from the archives of the brain. But I am a historian. I wanted to immerse myself in the Menninger archives to see what traces were left behind. I wanted to see records from 1946 to 1951, when we lived in Topeka. I wanted to catch my father—and his colleagues—in medias res, to find out what they were saying at the time they were saying it. If I could.

On my first day at the Menninger archives at the Kansas Historical Society, I located several files on a "Schizophrenia Research Project" containing correspondence, memos, and, most noteworthy, monthly reports detailing Dad's experimental treatment in real time with a patient whom he called "Nedda." The story was not quite as he had related it to us from memory.

They were an odd couple, Dad and Nedda, this Jewish psychoanalyst and former attorney from a middle-class background in Brooklyn and a Catholic working-class former telephone operator from a small town in Texas. He was forty years old when they met in 1948, married with two children, close to completing a PhD in psychology. Before coming to Menninger he had undergone a training analysis with Theodor Reik, an eminent colleague of Freud's in Vienna, and had attended Reik's seminars. Briefly a university professor and then a wartime psychological researcher in the U.S. Navy, he came to Topeka to take a position as acting chief of psychological services at the Winter VA Hospital. He also joined the staff of the Menninger Foundation just as Menninger was beginning its meteoric rise to prominence.

She was thirty-eight years old, Irish American, twice married, with no children and an education that had ended after one year of high school. Her mother had come from Ireland to the United States and wed a Texan from whom she was later divorced and who died while Nedda was in her early teens. Nedda's mother reportedly left much of the responsibility of raising her to Nedda's older sister, probably because, as a single mother, she needed to work to support her children, who also included two sons. The older sister, according to Nedda, resented such responsibility and treated her younger sibling harshly. Nedda married at the age of seventeen, divorced a few years later, and married again. In 1942, she enlisted in the Women's Army Corps but apparently never saw service. The following year she was hospitalized at Winter with a diagnosis of "hebephrenic schizophrenia," meaning the disorganized, noisy, delusional kind. She assaulted patients and staff alike and masturbated openly on the ward. She underwent many forms of therapy, including electroshock and insulin shock. But none of these approaches resulted in any improvement. Until my father took an interest in her and proposed to make her what he called a research patient, she seemed to be headed toward life on a locked back ward of the hospital, the fate of many of those with her dire diagnosis.[2]

Schizophrenia at that time was—and remains today—one of the most devastating and costly of all mental ills, "undoubtedly the major clinical and theoretical problem confronting psychiatry," according to one distinguished practitioner in 1950. Defined as a severe disorder of thought as well as of mood and behavior, schizophrenia was perhaps the most dreaded and most heartbreaking mental illness, for those with the disease and for their families. Not everyone diagnosed with schizophrenia experienced hallucinations and delusions or was completely divorced from their surroundings. But many were, at least part of the time. They often could not hold a job, live independently, or maintain meaningful friendships or intimate relationships. Many lived in fear, anguish, and confusion, haunted by terrors of being persecuted and having their minds and bodies invaded by hostile outside forces. To receive a diagnosis of schizophrenia in the late 1940s was often to be condemned to hopelessness and despair.[3]

Hospitalization offered some protection—people with schizophrenia occupied about one-fourth of all hospital beds in the United States in the late 1940s—including the possibility, in certain circumstances, of access to limited forms of treatment. As Joel Braslow, Jonathan Metzl, and others have recently argued, during the first half of the twentieth century, the idea that even the criminally insane might improve with treatment functioned as a viable concept; the goal of institutionalization was not containment but recuperation and return to the community. Certainly interventions could be harsh. At institutions such as Winter, electroshock and insulin shock therapy were widely practiced physical methods of treatment. Lobotomy, which had been developed in Europe in the 1930s, was also employed, much less widely but often with cruel results, leaving patients emotionally flat and cognitively impaired. But these methods often coexisted with occupational therapy, vocational rehabilitation, skills training, and even field trips. According to Braslow, "the 'total institution' of the 1950s was not nearly as harsh, totalizing, and dehumanizing as we might have believed."[4]

Relatively few psychoanalysts had attempted to treat patients with schizophrenia. One who did was Frieda Fromm-Reichmann, the influential and admired senior analyst at Chestnut Lodge, a private residential sanitarium outside Baltimore, Maryland, where patients received intensive, long-term, individualized treatment. And Karl Menninger too had long been interested in schizophrenia. He had treated a number of patients with that illness—sometimes using classic psychoanalytic methods—and promoted an attitude of optimism toward their therapy. It is understandable, then, that he took an interest in the controversial methods of a flamboyant Philadelphia psychiatrist, John Rosen, who in 1947 claimed to have cured a number of people of schizophrenia within a few weeks through his technique of "direct analysis" or "direct analytic therapy." Direct analysis meant responding to patients' pathological behavior by bombarding them with instant interpretations of their supposed unconscious motivations, such as "you want to fuck your father" or "you want to bite your mother's breast." Or as my father described it later, Rosen "intrudes himself on the psychotic with as much vehemence as we find in the statements of Fromm-Reichmann that one must never approach the schizophrenic without invitation." In Dad's words, "He educates, exhorts, interprets, lavishes affection, punishes, and participates in the psychotic reality in a way which certainly bewilders the cautious onlooker even if it does not dismay the psychotic."[5]

Rosen's claims and startling methods awakened Karl Menninger's curiosity but also his alarm: already Rosen was besieged by desperate families demanding access to this entirely untested treatment. So in early 1948, Menninger invited Rosen to Topeka to demonstrate his technique. Now my father also became intrigued. Years later, he described the demonstration as an abject failure. Rosen and the patient had sat in a room with a two-way mirror that allowed some of the senior analysts, residents, and staff, including my father, to observe their interactions. But in that setting the patient had completely ignored Rosen and his dramatic interpretations. Despite this disappointing outcome, at the time my father argued that by challenging some of the current concepts of treatment, Rosen had pointed to problems that were worth exploring systematically, with the hope that such exploration might produce hypotheses about therapeutic approaches that could be rigorously tested. In an era when antipsychotic medication was not yet available, when therapies were often highly physically invasive, when many people diagnosed with schizophrenia lived lives of desolation, Rosen's claims and approach were not easily ignored.

My father began seeing Nedda for several hours a day in June 1948 as an exploratory research project designed to test John Rosen's direct analysis techniques. A psychiatric resident attended every session, each of which was recorded.[6] My father also consulted weekly with the senior analyst on the project, the prickly, elusive William (Bill) Pious—considered one of the Menninger "luminaries"—who soon became his best friend. In November he began submitting monthly reports to the Menninger research department, headed by Sibylle Escalona, a child psychologist and enthusiastic supporter of the project. These early reports outlined the issues emerging in their daily sessions, especially Nedda's self-loathing and extreme guilt about sex, both in actions and in thoughts. My father indicated that she referred often to her strict Catholic upbringing and the guilt she felt because she had married a man who was not Catholic.

The experiment with direct analysis was brief. My father turned to less verbally aggressive, more supportive modes, though he was not ready to give up entirely on Rosen's methods. Three months into Nedda's treatment, he described his evolving approach as compelling his patient "to project all the internalized conflicts and problems on to him [the therapist] and thereby transfer them from the realm of inner fantasy to the outer and real object-filled world."[7] In this way, he wrote, "both because of constant interpretations and because I am consistently opposed to the irrational tendency uppermost at the moment, [she] makes me the protagonist in the great struggle with conscience and lust. She is no longer able to omit reality in the form of a real object as a factor in the problem."[8]

His monthly reports offered little description of Nedda's social background or ancestry, although as she began telling him more about her family he indicated that her two brothers had also been hospitalized with schizophrenia. Such similarities in a family history would later register as highly significant as genetics came to impinge on his experience, but here he mentioned it only in passing. In any case he was less interested in exploring the origins of Nedda's illness than in recounting the tenor of their daily interactions, which sometimes went on for two hours or more at a time. By November 1948, six months into Nedda's treatment, he was feeling "much more comfortable in the therapeutic situation." He believed that Nedda had achieved "significant advances" in her status, although he remained uncertain whether the project would "lead to valid hypotheses concerning the nature of schizophrenia and its treatment." Several months later he had not made progress in understanding "the schizophrenic process or its therapy" and was worried about the possibility of "some debilitating organic ailment complicating the schizophrenic picture."[9]

In March 1949, a new optimism entered his reports. "During this past month a considerable advance in the clinical status of the patient followed abruptly on a single variation in technique," he wrote. He did not explain what this variation was, but he was following up to see if the improvement could be maintained.[10] Four months later he was starting to believe that the hospital environment posed a serious impediment to Nedda's continued progress and "perhaps to the ultimate resolution of the psychosis," since he had so little control over her activities outside therapy hours. An enlightened ward doctor at the hospital helped him hire an aide to take her on outings in the afternoons and evenings. Thanks to this ward doctor, her "entire horizon seems to have widened amazingly." Nedda was swimming, going to shows, and even going to baseball games at Owl Park. The companion was "a good choice because [the] aide has no fear of the patient and seems to have picked up on my attitude quite readily." But now other problems intruded as Nedda was diagnosed with a large benign tumor that required a hysterectomy. After the operation she moved into a foster home in the community, a setting that my father felt was much better than the hospital and that, after an initial adjustment, was leading to further improvements in her mental state. By October 1949, he was feeling "more optimistic than ever," persuaded that "the transition from the hospital has helped very considerably."[11]

My father's monthly reports continued to emphasize the benefits to Nedda of her activities and her living situation outside the hospital and the need for continuing financial support to sustain her. She was becoming more communicative about her past life. By December 1949, he was considering that a visit to her home and husband in Texas might be beneficial, and the following month he described "steady progress" that far outstripped his "theoretical understanding of what takes place." After Nedda did make a home visit, in March 1950, he felt they were "clearly out of the first phase of combat and well into the second phase of interpretive psychotherapy." Still, he reported that uncertainties in the schedule made her extremely anxious. Visits home were challenging, especially since her husband seemed to be frightened by the prospect of assuming sole responsibility for his wife. Two months later she had "settled down considerably," was "more productive, more introspective, [and] more cooperative than at any [other] period of the treatment." Her foster parents, as he called them, who had worked with a skilled and sensitive social worker from Menninger, were pleased with the situation. My father, too, was feeling gratified and enthusiastic about the progress that Nedda was making. "I would hardly speak of her as being called 'cured' but I am quite happy to know that this much can be done for such a patient," he wrote in the report for May.[12]

In late June he presented the case of Nedda before the Topeka Psychoanalytic Society, to which he had recently been admitted. Now he explained the single variation in technique that he had alluded to in his report from more than a year earlier and gave a much fuller picture of the situation. Instead of trying to reassure Nedda that her sexual thoughts and feelings were normal, as he had been doing initially, he decided to agree with her self-accusatory, moralistic thoughts. He did not explain what had motivated this shift in strategy, but he began concurring with her that masturbation was wrong, that sex was bad, and that even thinking about sex was sinful. Perhaps even smoking cigarettes, as they were doing, was a sin. She seemed to grow calmer and more communicative with him, as if she were somehow reassured by his agreeing with her, even if he was agreeing with her own self-accusations. He felt that something in his "moralizing, something in this agreeable condemnation of frightening impulses, was responsible for the change." He decided to go further. He now began to forbid sexual and aggressive behavior and thoughts in a firm, even authoritarian, manner, encouraging her to get relief from prayer as she had before the onset of her illness. He also offered to pray for her. In subsequent therapy sessions, as his agreement with her moralizing continued to produce increasing periods of realistic thinking and thoughtfulness, he became more extreme. He sternly forbade "any sexual or aggressive manifestations which threatened to disturb the therapeutic relationship."

Nedda had always been physically assaultive on the hospital ward and during her therapy hours, though she had responded to verbal admonitions. But now, if she threatened to "use force against me," my father told her he would use "an equivalent force" against her. He noted that while her physical assaults were "rapidly diminishing" in frequency, they did still occur. And on those occasions,

"I did what I could to immobilize her. When the physical provocation was severe, I did more than inhibit her movements. I slapped her hard, sometimes very hard." In the unpublished manuscript of his talk, he had crossed out these last words, substituting instead, "I fulfilled my promise to use force with force," the words that ended up in the published paper. He claimed that his struggles with Nedda, with but one exception, ended "with peacemaking and affectionate exchanges of devotion." Only rarely did she fail to thank him for "putting an end to the threatening and overwhelming forces which had seized control of her."[13]

My father was not a violent man and only occasionally spanked my sister and me as children; he never slapped us, as far as I can recall. I never saw him use physical force against our mother or any other woman. His paper warned against interpreting his account as an endorsement of hitting or as the key to Nedda's improvement. But there were his words, acknowledging that he had slapped, repeatedly, his female patient. Reading them as his daughter, as a woman, I found his words distressing. Yet as a historian, I had to acknowledge that psychiatric hospitals in the 1940s and 1950s sometimes used harsh restraints to calm aggressive patients such as wet sheet packs, cold wet sheets that were wrapped around patients, and what the historian Jack Pressman has called "assaultive therapies": lobotomy, electroshock, insulin coma, sterilization, and the removal of teeth. Even that gentlest of therapists of schizophrenia, Frieda Fromm-Reichmann, sometimes saw assaultive patients in a wet sheet pack.[14] My father did not defend, much less endorse, the actions he reported with Nedda. But by acknowledging frankly, in his first formal psychoanalytic presentation and in his first professional publication the following year, that he had "met force with force," he raised questions and established a reputation, at Menninger and in the wider psychoanalytic world, that would trail him for years to come.

Over the summer of 1950, Nedda continued to make progress in almost every area, including her work at ceramics, work in her foster home, and social relationships with her foster family and their friends. "With greater and greater conviction," Dad wrote, "she reports her recollections of periods when she was not embarrassed by social relationships and by her thoughts. There are now quite sudden shifts in her whole psychological adjustment in which confusing and imperative thoughts seem to disappear and she is once again more 'real.'" She had periods when she seemed "altogether nonpsychotic. She herself reports sudden very marked changes when she returns to some former state when the world was clear and very close to her and quite understandable. These periods of clarity are not of very long duration, but she is able to remember them and report on them and to differentiate such periods from the peculiar clouding over which has been her constant state for the last eight years." My father believed that dealing with Nedda's anger and confusion in a firm, insistent way had led to the best therapeutic outcome.[15]

By September 1950, a little more than two years after they had begun therapy together, Dad considered that Nedda might soon be ready to return home permanently. "Except for certain rigid ways of thinking which on the surface appear psychotic and related to a severe thought disorder," he wrote, "the patient's intellectual and emotional behavior is quite appropriate. Although she seemed unable or unwilling to give up these ways of thinking, she was able to discuss her attitudes without getting extremely disturbed and losing contact with me and with her environment. Principally these problems relate to questions of dying and questions of sex." With the significant advances she had made, he believed it was likely that she would adjust to her home situation. But there were still "many psychotic residues," and he believed that the best assessment would be made when she returned to Menninger in about six months.[16]

Just before Nedda left Topeka to go home in November 1950, she agreed to be tested by both Dad and Sibylle Escalona. The results were dismaying. They "indicate the presence of a severe psychotic disorder, far more sharply delineated in the test battery than in the clinical appraisal." According to Escalona, "The content and form of [Nedda's] responses resemble those characteristically elicited from 'burned out' chronic schizophrenics." The entire record, in Escalona's view, represented "a valiant attempt to bring order out of chaos through the application of moral principles." Nedda's thought processes were often fluid. Her judgment as to what was realistic and what was not was "most uncertain. . . . It would seem that her understanding of physical and social reality is fragmentary and totally inadequate and that a moral view of nature provides such clues for reasonable or acceptable behavior as are at her disposal." The results of the Rorschach test "may be considered a classic schizophrenic production." The Bellevue test also revealed significant problems. Escalona concluded that "a great many of the patient's expressed attitudes do not correspond to her actual feelings and underlying beliefs" but represent views that Nedda had adopted from the outside. Escalona's prognosis for the future was pessimistic though she acknowledged that many people with schizophrenia make "a workable adjustment in noninstitutional environments," so predicting on the basis of test results did not merit much confidence. Escalona believed that "her genuine attempt to remain in contact with and comprehend outer reality is made by means of attitudes and beliefs so tenuously related to her inner experience (so little 'internalized') that the prospect for maintaining even a 'social adjustment' without continued thera-peutic support seems very poor, even in the simplest sort of environment."

On the positive side, Escalona found that, outside her areas of psychotic pre-occupation, Nedda was able to pursue rational trends of thought and that her integrated functioning appeared disrupted more by confusion and uncertainty than by panic or disturbing impulses. She was able to maintain contact with her examiner and showed a desire for kindness and communication.[17]

My father indicated that he would have preferred to keep Nedda in Topeka longer but that an illness in her foster family had made it necessary for her to leave their household, and he felt it was better for her to return home than to start over in a new foster home. He did not consider this the end of her treatment but more an experiment to determine what going home would mean to her. A few months later he reported that she seemed to be making a good adjustment.[18]

However controversial, the paper my father presented to the Topeka Psychoan-alytic Society that June was accepted for publication in the *International Jour-nal of Psychoanalysis* and got him invited to speak the following December at a conference at Yale University School of Medicine with Fromm-Reichmann and another eminent psychiatrist, Jerome Frank of Johns Hopkins. He was also invited to be a commentator at the midwinter meeting of the American Psycho-analytic Association, along with Fromm-Reichmann, on a paper by Kurt Eissler, one of the most distinguished of Freud's former associates. These were heady invitations for a novice who had treated only one patient with schizophrenia and had only recently been granted membership in the Topeka Psychoanalytic Soci-ety (which automatically accorded him entrée into the American Psychoanalytic Association and the International Psychoanalytical Association). Dad was one of the relatively few nonmedical analysts accepted at that time. The last days of 1950 must have felt to him like the close of a triumphal year, his annus mirabilis. Nancy and I, aged five and eight, had little inkling of these triumphs—or of the shattering news that our parents had received in September of that year. We did not know then or for a long time to come.[19]

My father always spoke of his work with Nedda as if it were a casual undertaking: the Menninger Foundation had paid him to do whatever research he liked for half his time as a staff member; for the other half, he would see patients at the clinic. As I read through the foundation's files, however, I realized that in fact the Schizophrenia Research Project was a major formal Menninger endeavor, albeit an unconventional one, to which Karl and Will Menninger and Sibylle Escalona were willing to devote considerable effort and resources.[20] Dad and Escalona exchanged numerous draft proposals as his treatment of Nedda, which they had initially considered an exploratory venture, expanded into the formal project, with its recorded and transcribed sessions, psychiatric observer, periodic psychological testing, and outside consultant, Pious. Meanwhile Escalona and both Karl and Will pursued grants from the Pepsodent Foundation, the National Committee for Mental Hygiene, the newly established National Institute of Mental Health, and the recently organized Lasker Foundation. Still in search of funding one year into the project, Will Menninger reported to members of the foundation that the need for contributions was "critical because our case is the first one in history where complete electric recordings were made of the psychological treatment of a chronic schizophrenic patient." Will was not given to hyperbole, so it must have been deeply gratifying to my father that Will concluded his appeal by characterizing the study as "one of the most important we have ever undertaken."[21]

Some of Will's and Escalona's enthusiasm may have rested on their confidence in my father and their sense of him as an unusual man. Escalona reported to Will, with perhaps a slight exaggeration, that my father had "an extraordinary combination of clinical research skills, which include didactic psychoanalytic training, training in research methods, and extensive clinical and research experience." To Dad she wrote that the staff was confident that his experience gave "much promise of intelligent and sensitive application to the problem."[22]

As she and my father defined it, the main focus of the project was theoretical: to advance understanding of the theory and therapy of schizophrenia. A Swiss psychiatrist, Eugen Bleuler, proposed the word *schizophrenia* in 1908 to acknowledge a fraying or breakdown in the mind's ability to integrate perception, cognition, and emotion. (Bleuler took issue with an earlier term, *dementia praecox*, introduced in the 1890s by the German psychiatrist Emil Kraepelin, arguing that this disorder did not always emerge early in life, as *praecox* implied, and was not strictly organic in origin.) By the late 1920s, *schizophrenia* had been adopted as the term of choice to characterize extreme disturbances of thought, affect, and behavior, including severe impairments in personal relationships and activities of daily living, not caused by any obvious organic condition. Still, Bleuler assumed the ultimate source of the illness to be an aberrant or otherwise abnormal brain but accepted that the wide-ranging symptoms might also be shaped by psychological and social experience. However in the United States, psychiatrists tended to fall into two distinct camps: those who considered schizophrenia a brain disease and focused on organic factors, including genetics, and those focused on the mind and psychological and environmental causes.[23]

Freud, trained as a neurologist, also assumed the existence of an underlying biological, likely hereditary, substrate of schizophrenia. But he had begun to articulate a psychological theory of the disease as early as 1894 in "The Defence Neuro-Psychoses," although he had not yet differentiated between neurosis and psychosis.[24] He made his fullest analysis in his 1911 discussion of Daniel Paul Schreber, a judge of one of Germany's high courts who experienced episodes of a startlingly imaginative psychosis. Schreber was never Freud's patient, and the two men never met. But Schreber's memoirs, published in 1903 and a best seller at the time, offered Freud an opportunity to elaborate a rather convoluted theory of paranoia. Freud hypothesized that Schreber's illness grew out of an internal conflict over what he called unconscious homosexuality, a view that was soon widely accepted among psychoanalysts, although others, including Bleuler, raised doubts early on.[25]

Freud also presented two other hypotheses that became foundational for later psychoanalytic theorizing about psychosis, including my father's. Freud proposed that the hallmark of schizophrenia was not the noisy, wild, "crazy" behaviors—the delusions and hallucinations—that were often considered its defining feature. Rather, it was the person's withdrawal from the external world, which was a silent process. He characterized this psychic withdrawal as a detachment of libido, that is, of psychic interest in and desire for people and things in the external (real) world. To make up for the lost world, the ego attempts to create another world by manufacturing hallucinations and delusions. Far from being defining aspects of the disease, these delusions, according to Freud, were really efforts at restitution, a move toward healing.[26]

A few years later Freud expanded these ideas, making distinctions between the neuroses, now the mainstay of psychoanalytic attention, and the psychoses, including schizophrenia (which he had previously termed "narcissistic neuroses"). He had theorized earlier that, in neuroses, libido becomes withdrawn from people (objects) in the real world and attached instead to objects in fantasy, including objects in the unconscious. But the person with a neurosis remains connected to external reality and typically wrestles with distressing conflicts between conscious aims and unconscious desires, that is, between the reality-oriented ego and the instinct-driven id. Now Freud argued that in schizophrenia, the libido withdrawn from objects in the real world "does not seek a new object, but retreats into the ego; that is to say, that here the object-cathexes [emotional energy invested in objects] are given up and a primitive objectless condition of narcissism is reestablished." (Elsewhere Freud put it this way: "Neurosis does not deny the existence of reality, it merely tries to ignore it; psychosis denies it and tries to substitute something else for it.")[27]

In short, two different animals.

Ultimately Freud was not much interested in treating psychosis, although he did sometimes refer psychotic patients to other analysts in Vienna and Berlin. He argued early on that people suffering from schizophrenia—and psychosis generally—were not amenable to psychoanalysis because they typically could not develop a transference—the analysand's tendency to project onto the analyst feelings and wishes belonging to an earlier relationship, most often with a parent.[28] But some of Freud's followers disagreed with his formulation. After all, a number of distinguished analysts had been using modified psychoanalytic methods for years with patients diagnosed with schizophrenia. They included Paul Federn in New York City, Harry Stack Sullivan of the William Alanson White Institute in Washington, DC, and of course Fromm-Reichmann. All of them, according to my father, attempted to woo their patients gently into a positive transference and viewed each one "as a creature of the utmost psychic fragility."[29] That was not Rosen's approach or my father's style at all.

Already in Topeka in 1948, with less than a decade of clinical experience, my father was developing the confident language, rhetorical style, and sense of mission that he would later use in a much different context in relation to a distinct disease. In one of his letters to funders a few months into the project, he expressed his dismay at the disjunction between the depth and urgency of the problem of schizophrenia and the tiny amount of research devoted to it. In what would become a hallmark of his style, he emphasized the challenge and risk of his project, appealing to his prospective donors' ambition and imagination. "I do not offer an experimental plan or a research methodology which insures positive or negative answers," he concluded. "What I would like most of all is to find that the almost insuperable difficulties of the problem make it all the more attractive to you so that I can always count on your interest and collaboration." From Stockbridge, David Rapaport—soon to become my father's greatest mentor and a trusted friend—wrote, "I love the unorthodoxy of it."[30]

"I always had a lot of, what shall I say, not gall, not even guts. But I would try anything," Dad told me in 1969. "Often I would try just because I was scared of it. . . . I had that kind of adventurous turn of mind that wouldn't turn down something because of anxiety. Rather it seemed to push me into doing the things that made me somewhat anxious. Not overwhelmingly, but if they made me somewhat anxious—it seemed a challenge that I might come out on top—I'd take a chance and do it." It was precisely the seemingly insuperable challenge that no doubt attracted my father, not only of working with a difficult patient with schizophrenia but of doing so under observation and within specific research constraints. It was also possibly the first time in his life that he had been able to pursue a major project he cared about deeply and with the enthusiasm and support of people he admired and respected. He would work to alleviate the suffering of one individual whom others had written off as hopeless while also trying to make theoretical and practical inroads into one of the greatest mental health problems in America. And he would see if he could do better than his more experienced colleagues who had worked with such patients for a long time. So this was a test not only of John Rosen and his methods but also of himself.[31]

2

Out of Brooklyn

It is difficult to live in Brooklyn without becoming a deep thinker.

—Clifton Fadiman

When my father was in his nineties and still living on his own, I began helping him pay bills for he could no longer see well enough to do so himself. He would put each day's mail into a large blue-and-white canvas tote bag that he kept in his bedroom, and once or twice a week we would tackle the bag together. It became a ritual, a moment of intimacy with my father that I grew to cherish. On certain nights when we had finished, he would sit back in his chair and reflect on the past, as if he were still questioning the shape of his life. He would express a sense of wonder that he had lived such an eventful life, what he sometimes called his accidental life. He would enumerate all the lucky encounters that had opened up unexpected opportunities, such as meeting, in the summer of 1946, David Rapaport, who urged him to go to Menninger's just at its moment of dramatic transformation. He also expressed the view that character is destiny. As a Freudian he believed that early childhood experience crucially shapes individual character, laying down patterns of behavior that often last a lifetime. And yet he believed that people could change. How could he be a therapist—a psychoanalyst—if not convinced that people could improve their ways of living?

Dad grew up in Brooklyn, though not in the poor and working-class Jewish immigrant Brooklyn neighborhoods made famous by his literary contemporaries. He had neither yearned to join the WASP intelligentsia like another Jewish

Brooklynite, the author and editor Clifton Fadiman, nor looked longingly at
the "American" streets like the critic Alfred Kazin.[1] Nor was his experience like
that of his high school classmate, the legendary Hollywood agent Irving "Swifty"
Lazar, who recalled that "in Brownsville, you were in combat training from the
moment you learned to walk."[2] The Wexlers lived everywhere in Brooklyn *but*
Brownsville, and Dad felt American from the start, unlike those Jews who, as the
writer Vivian Gornick put it, struggled "not to sound like one newly arrived to
the culture."[3] Later in life Dad impressed some people as more New England
than New York.[4]

What he most recalled from his childhood was moving to a better home every
two years—though always in Brooklyn—as his father, Nathan Wexler, moved
precariously up the economic ladder, with an occasional drop into temporary
poverty. Moving, transiency, and a lack of permanence dominated Dad's memories of his early years but in a positive way. In memory his father's pleasure in all
these moves carried over to him and shaped his character in ways that enabled
him to adapt easily to change.[5]

Though he grew up in Brooklyn, Dad was born in San Francisco. Sometime after the Great Earthquake of 1906, my grandfather gathered up his wife,
Mollie, and baby son, Henry, and traveled west across the country in search of
good construction jobs. It was in San Francisco, in a small wood-frame house at
51 Portola Street, that Dad was born on August 24, 1908.[6] According to Henry,
the rabbi predicted that Milton would be a great man because of his big nose.[7]
Years later, Milton and Henry, two psychoanalyst brothers, recalled a story about
Dad's birth that placed sibling rivalry at the center. When Nathan informed the
two-year-old Henry that he had a little brother, Henry allegedly cried, "Throw
him out the window!" and ran off down the street, telling the police officer who
found him that he was going to the theater.[8] Dad said this had been a funny story
in the family for years, although neither he nor Henry seemed to find it amusing.
Despite their early competitiveness and living on opposite coasts, the brothers
remained close. Henry became a psychiatrist and subsequently received his psychoanalytic training at Menninger at the same time that we were there. He would
move with his family in 1949 to New Haven, Connecticut, where he became one
of the founding members of the Western New England Psychoanalytic Institute
and a faculty member at the Yale University School of Medicine. But even with
his status as a respected analyst, Henry lacked Dad's self-assurance and ease with
change. He suffered more from fears and inhibitions, for instance fear of flying,
which limited his travel to conferences and meetings. He gave the impression to
some of his colleagues that he felt he hadn't received his due.[9]

In the end, Nathan's hopes for good construction jobs in San Francisco went
unrealized, though he evidently did well enough to buy a Stanley Steamer, a

stylish new automobile that came to a sad end one day, according to Dad, when he let it idle too long and part of the engine melted beyond repair. After two years, Nathan and Mollie decided to leave San Francisco, shuttling back and forth for several years between Brooklyn and Cleveland, Ohio, where Mollie's sister Rose lived with her husband, Abe Goldstein, a good-natured jeweler. Apparently Dad's ease with moving did not preclude deep attachments because many years later he recalled his sorrow at leaving Cleveland, where he had made many friends and had a beloved kindergarten teacher who had instilled in him an early enthusiasm for school and books.

Back in Brooklyn, the family lived near—and for one year with—Mollie's widowed mother, Golda Skolnitsky (Mollie's relatives appear in the census records variously as Skolnitsky, Skolnietsky and Skolnetsky, later shortened to Skolnick), the grandmother whom Dad remembered as a kind of wise woman in the neighborhood. Dad always spoke of "Baba Goldie" with affection and admiration and recalled how she had underlined passages in her bible that he himself would have chosen for their psychological insight. Hereditary or not, Golda's gift for psychological counseling in the community evidently set an example for her Wexler grandsons, all three of whom became psychologists.

FIGURE 2.1   Golda Skolnick and Mollie, Henry, Norma, and Milton Wexler, circa 1916.

Courtesy of Jane Stern.

My father liked living with his grandmother and was fond of all his mother's side of the family, whom he remembered as warm and friendly and "great feeders"—Mollie's brother Leo owned a grocery store with large pickle barrels that Dad always recalled with pleasure. But he disliked his paternal relatives, who were "rich and remote." Nathan's mother had died when Nathan was a toddler, and his father, Morris, soon remarried. They had arrived in Brooklyn in 1884 or 1885, from the Ukraine, home to many Yiddish-speaking Jews until the pogroms beginning in 1881 prompted many Jews to emigrate to the United States.[10] Morris Wexler set up shop in the Williamsburg neighborhood of Brooklyn, first as a joiner and then a builder, in partnership with another recent immigrant, Wolf Balleisen, with whom he was involved in constructing several apartment buildings.[11] The partners suffered a bankruptcy in 1907, which suggests prior financial troubles that may have been an incentive for my grandfather, who had been working with them as a plumber, to move to San Francisco. However, Morris evidently rebounded quickly, forming two other companies within a short period of time. Dad had few fond memories of these grandparents, who lived in a house that was "quite grand" and where everything was clean and pristine, including them.[12]

Nathan Wexler was twenty-three years old, with an eighth-grade education and living at home, when he married nineteen-year-old Mollie Skolnietsky in late 1905 or early 1906. He was an outgoing, good-natured, generous man with a zest for life and sense of bonhomie, a bon vivant according to Dad but lacking intellectual and social interests, "a Democrat from beginning to end," a man who never questioned things and who had a very concrete, practical mind. Not for him were the institutions of working-class Jewish immigrant life such as unions and protest marches, socialist or anarchist politics, and Yiddish theater. Nathan was solidly petit-bourgeois, ambitious to make money in real estate. Dad always described his father as "a Willy Loman character," inept in business; he never forgot having to share a small bungalow with a rough-speaking family on Coney Island when his father's income had plummeted, and they were poor for a while. But most of the time he was growing up, he enjoyed privileges not shared by his friends, such as not having to depend on summer jobs to get through school or on day jobs while attending classes at night. My grandfather paid for almost everything.

Dad had fond memories of his brothers as children, including friendly competition with Henry and feelings of protectiveness toward his baby brother, Murray, ten years younger than he was, whom he thought an amazingly beautiful child.

He felt more distant from his musically talented sister, Norma, four years his junior, who seems to have borne the brunt of Mollie's frustrations and unhappiness. In her photographs Mollie is a handsome young woman, slender, with dark lustrous hair and dimples. She probably had no more than a third-grade education, but Henry and Dad agreed that she was "smarter by far" than their father; it was she who had valued education for their children, which she had pushed Nathan to support.

Henry thought Nathan and Mollie were badly mismatched, each bringing out the worst in the other—Mollie's moralism and melancholia and Nathan's hot temper and migraines, which caused my father a lot of anxiety in his youth and may have helped motivate his interest in psychology. "He may have seemed a tyrant at times but his noisy tyranny was nothing compared to her silent one," Henry wrote years later. Both sons complained that their parents had taught them little "except of course the traditional things like obedience, honesty, cleanliness, etc., and not too rigorously at that. They were relatively lenient." While Dad's memories centered on the Skolnick generosity and warmth, Henry was struck by the emotional difficulties of several of Mollie's siblings. "What a heavily tainted family the maternal side is," he wrote Dad in one letter. "Why aren't we crazier? Was it the infusion of warm, human blood from Pop's side?"[13]

The marriage lasted more than fifty years, but Nathan had a lover for twenty-five of those years, a memory that haunted Mollie even after Nathan died. For years he did not come home Wednesday and Saturday nights, with "Mother knowing what was going on and waiting, saying she feared scandal but also admitting that there was little behind the facade [of the marriage]." According to Henry she was "chronically afraid," and though she was devastated by what he did, she also loved him. "But she admitted she shouldn't have married him, that he was a sport." When I consider her lack of education and skills and the stigma of divorce for women at the time, I imagine Mollie as a tragic figure whose intelligence and enthusiasm for learning in a later era might have made possible a more fulfilling life.[14] By the time Mollie and Nathan came to live near us in the early fifties Dad had begun treating them more as his patients than his parents; they were kind and loving to Nancy and me but remained somewhat distant from our lives.

For Dad, a specific moment loomed large in memories of his parents' marriage and how he felt about his own. When he was in his twenties Mollie came to him one day, begging him to talk to his father and urge him to end his affair. Knowing his father's hot temper but realizing that he sided with his mother completely in this situation, Dad felt that "I had to go head-to-head with him and resolve the issue at once, or the family structure would disintegrate." As Dad told it in his memoir, when he met with his father the following day, "I really came of age." Nathan did not deny the affair but asked his son to see the other woman to tell her it was over, which Dad agreed to do. "Within five minutes of talking with my father I knew that I was in charge," Dad wrote. "I had suddenly sensed the bluster in him. I even saw the relief in his face that someone in the family knew how things ought to be and had the inner strength to bring it about. From that time on, I became my parents' parent, and in some ways, the leader of the flock among all of my siblings."[15] Dad carried this memory of his father's irresponsibility and weakness with him for the rest of his life. He blamed Nathan not for adultery but for his unwillingness to face his mistress, tell her the truth, and end the relationship in a forthright and honest manner. For Dad, being a man meant facing up to hard truths and taking responsibility for one's actions. He reserved special contempt for weakness in men, by which he meant both irresponsibility and passivity. It also meant what he called "subservience" to women—being "mama's boys," allowing oneself to be "dominated" by one's wife or mother. This construction of male "weakness" was a central antifeminist theme of 1950s popular culture, of course, represented in such films as *Rebel Without a Cause* and in mainstream psychoanalysis. Dad's version was a little different. Haunted by the memory of his father, he was especially scornful of men who "chased skirts" all over town but were afraid to leave their wives and risk being on their own. He loved his father, but he was disappointed in him. He would later look for male figures whom he could respect, especially intellectually, among his teachers and colleagues and among his patients, too.

Dad recalled making his own friends and going his own way "from the very beginning." In his second or third year at Commercial High School (later called Alexander Hamilton High), he began to distinguish himself through debating, a popular competitive sport for boys in high schools and colleges in 1920s Brooklyn. He considered himself shy and not a natural public speaker, but he wrote later that debating was one way he could make himself visible even if he had to work hard at it. Debating societies had emerged in the United States in the mid-nineteenth century, and by the 1890s, they were staging intercollegiate debate competitions. Dad told us that in the 1920s, interscholastic debates were as popular as football games, with several thousand cheering, raucous fans in the audiences. Debaters had cachet, especially debate team captains, which my father eventually became. Local newspapers covered the debates, and officials such as municipal justices, election commissioners, and school principals acted as judges. In the *Brooklyn Daily Eagle* you could follow high school debates over whether compulsory unemployment insurance should be instituted and whether the time had come for the United States to grant independence to the Philippines. Dad got a big break in his junior year when the captain of his Commercial

FIGURE 2.2 "I was basically a shy person." Milton at fourteen or fifteen.

High debate team suffered a (temporary) "nervous breakdown." The coach chose my father to substitute that night, and his team won against the more prestigious Boys High. Dad's team went on to win the Brooklyn debating championship for the second consecutive year. In a group portrait published in the *Eagle*, a solemn seventeen-year-old Milton, dressed like a future banker, smiles somewhat smugly in the back row while the erstwhile captain holds an enormous brass winner's cup, the team's possession for that year.[16]

After that Dad raced about in a whirlwind of activity. He was not only on the debating team but also in the drama and literary societies. In February 1925, his senior year, he was again chosen captain of the debate team, which dominated the many debates of the Brooklyn Interscholastic Debating League that season. That is, until the final championship debate, an event attended by more than 2,500 students in the large auditorium of Eastern District High School. That night Dad's team lost. But he evidently did not dwell on their defeat. This early experience of both winning and losing debates no doubt contributed to his ability to deal well with defeats and setbacks. He would later champion the idea that those who succeeded in life and relationships were people who were not afraid of rejection and failure. Even though your enemies are quite willing to remember your failures, he told me, most people remember your successes, and if you count yourself a success, other people will, too. He had little patience for those who hid behind rationalizations and excuses for why they did not try.[17]

Despite the debate team's loss of the championship, Dad personally was on a winning streak, becoming both senior class president and valedictorian. The principal called him first "among the outstanding individual members of the class." At a public speaking contest that spring, he won a $250 scholarship to the college of his choice from the Forensic League of Greater New York, the first time this prestigious prize was given. When he graduated in June 1925—one of only 222 boys to graduate of the 1,065 who had enrolled four years earlier—the *Brooklyn Daily Eagle* published his picture and referred to him as "among the honor men of his class," noting that of all the awards given out, "the most important prize was received by Milton Wexler," a gold watch for "general excellence." This prize, awarded by the Ben Franklin Club, may have helped inspire his affection for Franklin whose methods for moral self-improvement he would later turn to as a model for reshaping one's emotional life. Dad looked on his honors with a certain skepticism, however, for his grades were never stellar. As he put it wryly years afterward, "I went out in a blaze of glory," with a status he seemed to feel he did not deserve.[18]

CLASS PRESIDENT WILL
ENTER HARVARD IN FALL

MILTON WEXLER

Milton Wexler is president of the senior class at Alexander Hamilton High School which held its commencement last night. His name is among the honor men of his class and he has won laurels for his school on the debating team. He also won the Forensic League scholarship offered by New York University. He addressed his class last night.

FIGURE 2.3 *Brooklyn Daily Eagle,* June 25, 1925.

For all his later boldness, Dad's valedictory speech was surprisingly conventional, praising "patient effort, industrious application, and the power to see things through." But though his phrases were familiar, glimmers of the thoughtful and questioning man he would become could be seen, as when he stressed the "capacity to see something in life besides its matter" and when he urged his classmates to pay attention to "the spiritual, the finer side of life." He encouraged them to " 'look for the invisible' for the things not seen are eternal."[19] He did not mean God or religion, which he regarded with growing hostility; indeed he liked to tell the story of how, the day after his bar mitzvah, he announced to his parents that he was never going to enter a synagogue again unless for a friend's wedding or funeral. In speaking of the spiritual, I think he was talking back to his father, speaking against what he perceived as Nathan's mundane and material interests. He wanted something more, but he did not know exactly what.

Later in his life, my father liked to look back on his young self as a budding scientist. He recalled his childhood interest in radio, which had stirred an early fascination with science. Radio technology was developed in the early 1900s, and radios became popular following the sinking of the *Titanic* in 1912 and especially during World War I. A childhood friend got Dad interested in building crystal radio sets, and Dad recalled impressing his startled parents with the fact that, from their home in Brooklyn, he could listen on his headphones to stations in Pittsburgh and Schenectady. "I was experimentally minded," he told me in 1969. He felt that building radio sets also gave him an early sense of confidence and mastery. "It set me apart as a scientist who was in on the early elements of physics and radio wave transmission," he said. But Dad may have projected his later interest in science back into the past, for his fascination with radio did not inspire him to pursue science courses in school or even hobbies outside of it. The only undergraduate science course he took was botany, the only graduate class a required course in neuroanatomy. Psychology, political science, and literature were more to his liking. Especially literature, reading novels. The varieties of human character fascinated him far more at that point than those of the natural world.[20]

Dad dreamed of going to Harvard, the college he perceived as the pinnacle of success. But as he told it later, his brother Henry and his cousin Mort Starobin, a star college football player, persuaded him to join them at Syracuse University instead; he was not as independent as he later imagined. He regretted that choice afterward, but had he gone to Harvard, he might have experienced the wrenching anti-Semitism that Delmore Schwartz described, recalling how much he had suffered as a Jew at Harvard when he had taught there during the Depression. As Vivian Gornick tells it, the patricians of the Harvard English Department recoiled from Schwartz. He found himself both aggrieved by their rejection and yearning for their recognition and acceptance, for which he hated himself. By attending Syracuse with its greater Jewish student population, Dad escaped the deeply wounding encounters that shaped so much of the anger and self-doubt many young Jewish intellectuals suffered during that time.[21] In any case, Dad ended up attending Syracuse for only two years, 1925 to 1927, the minimum required to enter law school, although he managed to cram in three years of course work with the comparative literature classes he took over the summer of 1926 at Teachers College of Columbia University, his introduction to the school he would return to thirteen years later. He earned As in English, rhetoric, and political science and Bs and Cs in everything else. He had athletic, more than academic, ambitions, trying baseball, football, boxing, long-distance running, and even fencing, his brother Henry's sport, without much success. He still excelled in debating, however, which kept him out of class and traveling too much. He was just "seizing on opportunities as they floated by me," he said. He had no plan for his life and no mentor with whom to discuss options.[22]

But Dad was already acting as a kind of mentor to others. Years later he recalled doing informal therapy during his college years, when friends would often come to him with their problems. Besides the influence of his grandmother Goldie, he connected his youthful abilities as a counselor to his role one summer as caregiver to his brother Murray. When Murray was just three or four, he had an operation for mastoiditis, a type of ear infection that was quite serious in the 1920s. Dad was about thirteen or fourteen at the time, the older Henry was away from home, and their sister Norma too young to take on such responsibility. So it was Dad who became Murray's caregiver. He adored his little brother and liked carrying him around on his back to baseball games and giving him a lot of attention. He later believed this experience encouraged his therapeutic interests. He had a knack "for bringing out problems that people had, for being a counselor," he told Nancy and me. "It was a kind of role." He also acted as a therapist for his college roommate who was very shy and obsessive. And for a year he practically operated as an analyst to another student, Michael Dunn, who became an influential and lifelong friend. According to Dad, Mike talked to him about all his problems, regarding Dad as the more sophisticated city boy while he was the naive country boy. "Which wasn't true at all," Dad said, but that, too, was "a kind of counseling, therapeutic role."[23]

Since Henry was in the pre-med program at Syracuse and since Dad excelled at debating and was attending college on a public speaking scholarship, he felt that law was his logical destination, even if he had no passion for that vocation. In the stories he told late in his life, he portrayed his youthful self as one who simply took the easy road ahead of him and did not really know what he wanted. Both he and Henry believed that they had lacked essential parental guidance in their youth and that their educations had suffered as a result—a theme to which

FIGURE 2.4 At Syracuse University. Milton standing at left, circa 1925 to 1927.

Courtesy of Jane Stern.

Dad often returned. He would later go to great lengths to advise and mentor Nancy and me, even embarrassing me on one occasion by writing, unbidden, to one of my graduate school professors protesting a low grade he thought unfair.

In the fall of 1927, he was admitted to the New York University law school where he chose to attend classes at night so he could be with his friends who had jobs during the day. He had just turned nineteen and continued to live at home with his parents, Murray, and Norma (Henry was gone by this time) in their single-family house on Ocean Parkway Boulevard in the comfortable Midwood section of Brooklyn, surrounded by neighbors who included lawyers, clothing manufacturers, and contractors. By the mid-1920s, Nathan was flourishing as a real estate developer, so much so that the family could spend entire summers vacationing in the Catskills, an option not available to most of Dad's friends.

At this point in his life, Dad preferred to be near those friends, so he elected to take summer jobs that he chose on the basis of geography. One of the most memorable involved taking care of a young man about his own age who had been hospitalized in a New Jersey sanitarium diagnosed with paranoid schizophrenia. Dad lived in the sanitarium with this patient for a while, unaware that he was potentially dangerous in his delusions and extreme paranoia. Instead of being put off, my father found the patient interesting and wanted to learn more about the illness. He also did informal psychotherapy with another young patient at the sanitarium who supposedly had an advanced case of Parkinson's disease and, in my father's memory, had looked upon him as a savior. These experiences drew him even closer to Mike Dunn, who was now studying psychology, and to the circle of psychologists around Dunn and his wife, Mildred. Dad found their discussions far more compelling than the law.[24]

With his days free during the school year, he read a lot of novels, especially those of nineteenth-century Russian writers, some of whose works had been newly translated into English, writers such as Dostoyevsky, Turgenev, Tolstoy, Gogol, Pushkin, "and any other Russian author I could get my hands on." He felt that all these writers mixed well with other authors he was reading, including Havelock Ellis, Frank Harris, and especially Freud, whose image had appeared on the cover of *Time* magazine in 1924. According to most historians, it was in the 1920s that American interest in Freud spread beyond the small circles of Greenwich Village bohemians and other avatars of the avant-garde who were the earliest Freudian enthusiasts in America, drawn to psychoanalysis primarily as a path to personal and sexual liberation.[25]

Dad was too conventional to be a bohemian. He was not countercultural, but he was curious. He found that this new way of thinking resonated with the fiction he was reading and with his experience at the sanitarium. But he could not imagine a professional path that drew on these interests. He could not conceive of becoming a psychoanalyst. He had no interest in medical school, increasingly the path toward becoming an analyst, and could not envision—nor perhaps afford—traveling to study with Freud or with his students or associates in Vienna or elsewhere, as some affluent young Americans were doing.[26] Clinical psychology as a field barely existed, and most psychologists at that time performed diagnostic and intelligence testing or experimented on rats in university laboratories. It was as if his speaking and debating skills tethered him to a career in law, whether he liked it or not.

What remained most vivid in my father's memory from his three years in
law school was his boredom—also coaching his friends and getting booted out
of one of his courses by a famous judge for reading Freud in class.[27] But despite
what he described as his lackadaisical approach to studying, he managed to get
the highest grades of all the freshmen law school class and was appointed to

FIGURE 2.5   "I was good at writing briefs. But I was never proud of it because it always seemed
to me a trick. I could write a brief without any conviction whatsoever about the justice of the
cause." Jan. 1938.

the prestigious *Law Review*, where he served as decisions editor. By the time he passed the bar at the age of twenty-one, in 1930—one of 851 who passed out of 2,219 who took the exam that year in New York—the country was heading into the Great Depression, with millions out of work and unemployment soaring. It was not a great time for embarking on a career as an attorney, or any career. He took the first clerkship offered to him and eventually opened his own small law firm at 42nd Street and Broadway in Manhattan with Henry Sternberg, a friend of his brother Henry's, who would remain Dad's lifelong friend as well and who, along with his daughters who had muscular dystrophy, taught him about the challenges of having a hereditary illness in the family long before he discovered one in his own.[28]

In later years Dad would contrast his sense of drift at the time with the passion of a young man employed in their law office. Jerome Weidman, five years younger than my father, was writing a novel, which was later published as *I Can Get It for You Wholesale* and became a best seller. He knew he wanted to be a writer and took a day job as an accountant to support his writing. Observing Weidman's dedication only deepened my father's awareness of his lack of interest in the law. Many years later, however, he reflected that, though he had found his legal experience dull and alienating, he had also found it useful. Thinking like a lawyer was "so very opposite from the loose, intuitive, free associative ways that characterize psychoanalysis," he wrote. "Having had to think both ways has proved of real value. . . . Certainly I've been able to communicate with a much wider variety of people as a result."[29]

One of the people Dad communicated with while in law school left an enduring memory. He was working as a bellhop and elevator operator one summer at a New Jersey resort and had many conversations with a guest who invited him to go on walks and engaged him in long philosophical conversations. That guest turned out to be the head of the Jewish Theological Seminary. Before leaving the resort, he told my father that he should become a rabbi " 'because you can reason. Because you are thoughtful. Because you have an interest in abstract things and common human preoccupations. Because you speak well.' " Although he had no interest in entering the rabbinate, those words must have impressed my atheist elevator operator father because fifty years later he recorded them verbatim—or his recollections of them—in his memoir. And in a way, he did become a rabbi, albeit one who did not believe in God.[30]

3

Becoming Freudian

So many in number are the things that we experience without knowing them, and the profound inner realities that remain hidden from us.

—Marcel Proust, *In the Shadow of Young Girls in Flower*

My father was still a practicing attorney, a "suit" he called himself years later, "real buttoned up," when he met Leonore Sabin at a party in late 1935. She was a graduate student, six years his junior, in the illustrious Columbia University zoology department made famous by the fruit fly geneticist Thomas Hunt Morgan, who had won a Nobel Prize two years earlier for elucidating the role chromosomes play in heredity. By the time Mom arrived, Morgan had moved to the California Institute of Technology in Pasadena, but the Columbia department remained the most distinguished in the country, and Mom earned As and Bs in such graduate courses as advanced genetics, cytology, comparative embryology of vertebrates, and organic chemistry. When she met Dad she was finishing a master's thesis on the genetics of drosophila at a time when genetics was a relatively new academic field.[1]

Mom had been an outstanding undergraduate at Hunter College, which she had attended on a scholarship, majoring in biology in an era when few women entered science. She won academic honors, one of which enabled her to spend the summer of 1933, between her junior and senior years, studying entomology at the University of Michigan Biological Station in Cheboygan, Michigan, on the shores of Lake Huron. When she graduated Hunter cum laude, she won the

prestigious Elsie Seringhaus Prize, an award given to the student with the highest academic achievement in biology who also participated in extracurricular
activities in the department (she was president of the Biology Club). This prize
enabled her to spend the summer of 1934 at the renowned Marine Biological
Laboratory in Woods Hole, Massachusetts, where she took another entomology course. Unlike Dad, she had held a job while going to school, at least on
Saturdays, when she went downtown to work at Macy's department store.[2] She
was living in an apartment on West 78th Street in Manhattan with her widowed
mother, Sabina, and two older musician brothers, one of whom had a modestly
successful band. The Paul Sabin Orchestra played at such well-known New York
City nightspots as Tavern on the Green in Central Park, a gig of which Mom
was especially proud. A third brother, who worked in sales, was married and
lived apart.

As he described their romance years later, Dad was less entranced with Mom
than with her lively intellectual environment and her witty, fun-loving friends.
In memory at least, he was drawn to her more as an escape from his tedious life
as a lawyer than as a woman he loved. They planned a small wedding in the fall
of 1936. But Mom kept getting sick, leading to delay after delay, which my father,
in retrospect, felt should have awakened questions in his mind. He knew her
father had died a few years earlier, but he never thought to ask the cause. Even
if she had told him, as she claimed she had, he might not have inquired further.
In fact, as she well knew, Abraham Sabin had died of Huntington's chorea—
one of the most dreaded hereditary neurological disorders, well known among
neurologists and geneticists at the time, though not among physicians generally. Huntington's chorea was typically an illness that developed in midlife,
characterized by steadily increasing involuntary movements all over the body
(chorea) along with personality changes and a decline in cognitive abilities. It
progressed steadily without remission, stealing capacities for work and joy and
leading inexorably to death, usually over a period of ten to twenty years. Depression and extreme irritability were common. Geneticists held it up as a classic
example of autosomal-dominant inheritance, meaning that each offspring of an
affected parent had a 50 percent chance of inheriting the disease if they lived
long enough. And Huntington's had a long history of stigma, being considered
one of the worst hereditary ills by early twentieth-century supporters of eugenics, who called for sterilization of those living with or vulnerable to the disorder.
In the 1933 Law for the Prevention of Hereditarily Diseased Progeny, the Nazi
regime in Germany included people with Huntington's among those targeted
for compulsory sterilization. Ordinances prohibiting marriage by those considered unfit soon followed.[3]

These developments were not kept secret. The Nazis were inspired in part by the eugenics movement in the United States, and widely publicized communication went back and forth between Germans and Americans on the subject of eugenics and "race hygiene." The U.S. press, such as the *New York Times*, followed German sterilization efforts closely, sometimes with approval, reporting on the expansion of the law and debates within Germany.[4] By the 1930s many American states had enacted laws legalizing the forced sterilization of those in mental institutions for such vaguely defined conditions as "feeblemindedness." By the 1970s, approximately sixty thousand individuals who landed in institutions for one reason or another, a majority of them women, had been sterilized against their will and often without their knowledge.[5]

My mother and father could not have avoided reading in the press about both German and American eugenic sterilization debates and laws. They likely would have known about the German law passed in March 1935 that limited marriage to "the hereditarily healthy." Mom would have known that any knowledgeable physician or social worker, had she asked, would have advised her against marriage or strongly recommended that she not have children and preferably undergo sterilization. People like her, with a severe hereditary disease in the family, were often referred to in the press as defective, undesirable, and unfit. Moreover, these were not simply the views of reactionaries. Eugenics was generally considered modern and progressive; even intellectuals and scientists on the left, such as the future Nobel laureate Hermann Muller, defended the use of modern science to "improve" society through the control of reproduction although, by the 1930s, some university geneticists and other academic scientists had begun to criticize eugenics as scientifically and socially naive as well as racist. But most did not oppose the goals of eugenics. Rather, they sought to reform the means.[6] Even respected scientific publications that Mom likely read, such as the *Journal of Heredity*, praised the Nazi sterilization law as "proceeding toward a policy that will accord with the best thought of eugenists [*sic*] in all civilized countries."[7]

No wonder, I think now, that Mom kept getting sick and postponing the wedding. No wonder that among affected families in the United States, Huntington's chorea was shrouded in secrecy and shame. Medical literature sometimes cast members of these families who had children, despite knowing the risk of inheritance, as selfish and irresponsible, as if lives with a risk of Huntington's were not worth living. At the same time, those who did wish to avoid childbearing found birth control difficult to access and abortion a criminal act. Today I can only try to imagine my mother's anxiety and fear—or perhaps denial—as she contemplated marriage to my father while reading about proposals or legislation to sterilize or eliminate the "unfit." Why would she, a student of genetics, even

FIGURE 3.1 "There was no great romance." Milton and Leonore, circa 1937.

consider telling her fiancé about Huntington's chorea if she wanted to marry and have a family? And why shouldn't she pursue what was central to the identity and desires of most young women of her generation? The illness would have seemed remote, not appearing until far in the future, and the risk was only fifty-fifty. But if Milton Wexler had discovered what the doctors and scientists were saying about Huntington's, would he have married Leonore Sabin? I doubt it.

Years later Henry claimed that Dad did tell him about the presence of Huntington's in the Sabin family and that he had advised Dad against the marriage. My father, on the other hand, insisted that he had not known and had no memory of such a conversation with his brother, though he chastised himself years later for not asking about the cause of his father-in-law's death. Whatever the truth of the matter, Leonore and Milton finally married on October 2, 1936, in a small ceremony at a hotel on the Upper West Side of Manhattan. Shortly afterward they moved into an apartment at 183rd Street and Broadway, in the Bronx. Dad was still practicing law on 42nd Street. Mom completed her M.A. in zoology at Columbia and found a job as a substitute junior high school biology teacher. Soon she was hired to teach biology at Walton High, a large public high school for girls in the Bronx, the only candidate, she noted years later, to receive a 100 percent rating on the oral and interview qualifying tests.[8] And so, with the opening battles of revolution and civil war in Spain marking the front lines of World War II, my twenty-eight-year-old father and twenty-two-year-old mother with a devastating family secret settled into their married life in New York City, with two jobs and a decent income, surrounded by family and new friends.

My father liked to recount how, in 1938, he decided all at once to quit his law practice, return to school to study psychology, and begin a training analysis with Theodor Reik, a former student and colleague of Freud's in Vienna who had arrived in New York as a refugee from Naziism in June. I like to think that his marriage to my mother began this seismic shift.[9] Stepping into her world of intellectuals was his first move away from the conventionality and drift of his life, enabling him to take the plunge into psychoanalysis even if the summer of 1938, in the depths of the Depression and with Naziism spreading its tentacles across Europe, did not seem the ideal moment for a young married Jewish man to begin lying on a couch several times a week in a small New York office spinning out free associations and talking about his dreams.

By 1938, Freud had become internationally famous, recognized as a great thinker and writer whose new publications and English translations the *New York Times* regularly reviewed.[10] A few psychoanalysts were now practicing in the United States—by 1941 there would be one hundred, the majority located in New York.[11] However, as more refugee analysts began arriving from war-torn Europe, the New York analysts became less welcoming to the newcomers, whom they regarded not so much as distinguished colleagues but more as professional competition. Eager to fend off hostility to psychoanalysis from conservative medical quarters in the United States on the one hand and wary of ill-qualified Freudian enthusiasts on the other, the American Psychoanalytic Association passed the so-called 1938 rule decreeing that, except for a handful of prominent leaders trained in Europe before 1938, only psychiatrists—that is, MDs—were eligible for psychoanalytic training and membership.[12] PhDs did not qualify—this despite the fact that early European psychoanalysis had drawn followers from fields outside medicine. But even in Europe, younger analysts were aligning themselves more and more with psychiatry and medicine as a move for greater professional legitimacy. Lay analysts, as nonmedical practitioners were called, faced increasing barriers to acceptance on both sides of the Atlantic.

Reik was a lay analyst with a PhD in psychology and a literary bent—he had written a psychoanalytic interpretation of Flaubert's novella *The Temptation of Saint Anthony* for his dissertation—who came to New York with impressive credentials.[13] He had been a member of Freud's inner circle and attended Freud's Wednesday night gatherings at his home. Freud had famously come to his defense in 1927 when an American patient of Reik's in Vienna sued him, claiming he was an illegitimate practitioner because he lacked a medical degree. The case won wide publicity, including in the *New York Times*, where my father no doubt read about it.[14] Freud's defense, published as *The Question of Lay Analysis*, is still one of the best popular explications of psychoanalysis available. The essay laid out Freud's conviction that, in addition to undergoing a personal analysis, a broad education in the humanities—which he had encouraged Reik to pursue— was the best preparation for becoming an analyst. Freud strongly opposed the concept of psychoanalysis as a handmaiden of psychiatry or general medicine, arguing that it was a theory of psychology and should remain an independent discipline. The lawsuit against Reik was dropped.[15]

Reik's arrival in New York, then, ignited my father's imagination, allowing him to consider that he might become an analyst, too. He envisioned Reik as someone he could emulate, an analyst with a degree that was within his reach. And if he could enter a personal analysis with Reik, the most important prerequisite for becoming an analyst, he felt he could master all the rest. But he faced an immediate practical obstacle: he did not even have a bachelor's degree, having qualified for law school after just two years as an undergraduate at Syracuse, and a mediocre undergraduate at that. Would any university accept him? he wondered. He was ecstatic when, in the fall of 1938, he was accepted into the graduate program at Teachers College, where he could first take the requisite courses for a BS and an MA and then pursue a PhD in educational psychology. Teachers College also offered clinical courses and experience, being one of the few universities to include such training; for the most part, academic psychology at the time meant conducting laboratory experiments or intelligence testing, not seeing patients in an office or clinic.

To leave a law practice to become a student and, at the same time, undergo an analysis, perhaps even to contemplate taking such bold steps, one needed not only imagination but an income. How were he and my mother to live, especially in the middle of the Great Depression? Willy Loman or not, my developer grandfather must have been prospering to have bestowed on my father an extremely generous wedding gift that made my father's dreams even conceivable. At the time of Dad's marriage two years earlier, Nathan had given him an empty lot in New Jersey on land where he was building homes. With cash from the sale of that lot and practical assistance from his sister Norma's husband, my father was able to buy an entire apartment building in the Bronx, an astounding purchase for a young lawyer at that historical moment. Equally astonishing, he then had it renovated and rented out some seventeen apartments for $25 or $30 a month, yielding around $2,000 a year—close to the average American annual income in 1938.[16] Adding to this bounty, Henry Sternberg agreed to give him half the proceeds—as they came in—from all the pending legal cases that my father had worked on thus far. And my mother had an income from her job teaching as well. Compared to most of their peers, they were well off.

Dad always talked about attending Columbia University; his PhD was from Columbia. So for a long time I did not pay attention to the fact that he attended Teachers College, a school of education with a distinct identity within the university. In the era before clinical psychology became well established as a discipline, Teachers College was not only one of the few academic institutions offering clinical training; it was also the most progressive, innovative, and influential teacher training school in the country, one that placed emphasis on community involvement and social change. The atmosphere during the 1930s, according to former president Lawrence Cremin, was "electric," "charged with interest and excitement." The founding principles of Teachers College in the late nineteenth century continued to animate its culture in the thirties: that education should be active, not passive; that education for democracy should have a social rather than an individualistic focus; and that education should encourage creativity and experimentation rather than imitation. While Teachers College emphasized the training of useful citizens, it was also committed to supporting rigorous research, both in educational methods and in specific disciplinary fields. One of the touchstones of the college was a famous yearlong interdisciplinary course instituted in the early 1930s, Educational Foundations, or Educ. 200, which emphasized a critical approach to the study of the social world, particularly to notions of American individualism. It promoted a recognition of the interdependent nature of society and of the entire world and the value of human rights over property rights.[17]

Dad took that course. I found it on his transcript, along with a lot of education classes and a few psychology and neuroanatomy courses sprinkled in. With his laser focus on psychoanalysis, Dad did not pay much attention to the critical social perspective of Teachers College. Yet I believe he was influenced more than he acknowledged by the progressive and innovative culture of the place, carrying

it with him into his later work at Menninger and in Los Angeles, even if he did
not make the connection.

He described his time at Teachers College in different ways at distinct
moments of his life and depending upon whom he was addressing. "I went
through agony in grad school because very little of it appealed to me at all and
I had my eye on psychoanalysis, a subject almost never touched on in school,"
he wrote to Nancy and me in 1965 when we were both struggling in graduate
school, and he was trying to give us encouragement and reassurance. "But I knew
I needed the PhD; I knew I needed it in clinical psych; and I was willing to swallow anything to reach the point that would allow me to operate in my own frame
of reference with the proper credentials."[18] Later he presented a more positive
view, emphasizing that once enrolled he began meeting interesting people, and
things came alive for him, so that he found the training both exhilarating and
informative. He especially enjoyed the clinic (Guidance Laboratory) experience
in which graduate students practiced their clinical skills and discussed their cases
with such faculty members as Bruno Klopfer, a pioneer of Rorschach diagnostic
testing, and Robert Challman, an educational psychologist who later came to
Menninger, two professors who became friends.[19]

He remained critical, however, of the overt hostility toward psychoanalysis
that permeated the faculty even as many professors were secretly intrigued. He
also considered his faculty advisor, Percival M. Symonds, head of the Department
of Guidance, "the dullest man alive." He couldn't bear the constant contact
with a man he considered rigid and constricted. However, Symonds had been
extremely kind and generous to Dad, helping him get accepted at the college
and being supportive throughout Dad's sojourn there. He was even interested in
psychoanalysis, although too timid to pursue the subject at the time. And Dad's
work as Symonds's research assistant (from September 1940 to June 1941) developing projective techniques for analyzing the fantasies of adolescents gave him
his first practice doing serious research in psychology, a valuable experience that
would later help qualify him for research positions during World War II and
afterward in Topeka. It also became the basis for his first publication as a budding
psychotherapist.[20]

"The Story of Frank" bears all the hallmarks of my father's later psychoanalytic writing, especially his gift for hearing the pain and hurt beneath the hostile,
sullen, or simply silent surface of his interlocutor and his pragmatic, down-to-
earth approach to problem-solving. As part of his research, for five or six months
he had engaged in a lot of "plain and fancy 'bull sessioning'" with a fifteen-year-
old "bad boy" he called Frank (a pseudonym). From Frank's stories, my student
father concluded that "here is a yearning for affection and understanding. Here is

a denial and rejection of wrongdoing. What would it take to translate such a fantasy into action?" Despite the fact that he was not aiming to "cure" Frank, my father gathered Frank's teachers together in a case conference to discuss Frank's possibilities, and from that meeting the teachers began to talk to Frank differently. They approached him in the halls and after class "not about good and bad, sin and morality, but about his dog, swimming, teams, school activities." When he slipped, they talked it over with him. His foster mother took him swimming, his machinist foster father took him to visit his machine shop. Although Frank never became a model student, he became less hostile and graduated from junior high. Frank ended up serving in the U.S. Navy. My father left his readers to draw their own conclusions "about the role of the school and the home in the adjustment of exceptional pupils."[21]

Dad never spoke to me about this case or wrote about this research project beyond dismissing it as uninspired. But after he died I discovered his copy of Symonds's book *Adolescent Fantasy: An Investigation of the Picture-Study Method of Personality Study*, published in 1949 and sent to him "with the compliments of the author." The book revealed that my father was one of only two graduate research assistants on this major project and was given considerable leeway for making interpretations. A lengthy chapter, "Case of Jack," was undoubtedly Dad's story of Frank, in which he urged Jack's teachers to adopt a more sympathetic and understanding attitude toward the boy. "Jack needs a firm, but kind hand," Dad wrote, not the harsh and punitive responses he had received most of his life.[22]

Dad's case report of Frank/Jack was so like my father's later writing, written with his characteristic rhythms and with his intuitive grasp of the teenager's sense of rejection and the longing for affection beneath his sneer. Clearly Percival Symonds opened doors for my father, even if he could hardly hold a candle to the swashbuckling maestro from Vienna, Theodor Reik. For all Dad's criticisms of his adviser's timidity and lack of intellectual adventurousness, Symonds became an influential educational psychologist who later incorporated many ideas from psychoanalysis into his psychology textbooks and studies of personality. In his three-volume *Dynamics of Psychotherapy*, which Dad also carried with him through all his many moves, Symonds explained his point of view as eclectic and also acknowledged a major debt to Freud. He comes across as humane and progressive in his acknowledgement of the stigma surrounding psychotherapy, which "most persons still think of as 'a blight' and an admission of weakness." He argued instead that a newer point of view was emerging that made psychotherapy an opportunity to overcome one's limitations and make the most of one's abilities. Far from turning people into mediocrities and destroying their creativity, psychotherapy "unlooses creative talents" and can help people become more comfortable with themselves, Symonds argued. In his view, the growing demand for psychotherapy presented a golden opportunity for many young people to pursue training for this promising profession.[23]

My father managed to get through his courses at Teachers College without too much difficulty. But equally important to him was a renewed feeling of confidence and self-esteem, which his law practice had diminished. As an older student and former attorney, with a "reasonably good" reading knowledge of Freud, he found himself respected and admired by his fellow students, even surrounded by a certain prestige. He felt that his ability to express himself "in ways that were, in a sense, foreclosed to them" also helped him establish himself.[24]

Later Dad summed it up this way:

> So all at once I quit law, began my analysis with Reik. And you know, the way Reik worked, you went in one day, and the next day you were on the couch. So I was in analytic training, and I was admitted to Columbia as a student in clinical psych. And there was that marvelous transformation that took place almost overnight. And of course it meant that it exposed me to many contacts that up to that point I had been completely deprived of. And I must say that now, mixing with the university students, with the professors, and being analyzed, which somehow or other, although Columbia pretended that it would have nothing to do with analysis, made me very prestigious. In those days the professors wouldn't dare go into analysis, but they envied all those people who were in analysis. And to think that I was in a training analysis was something that they secretly envied tremendously. So that I pretended to keep it a secret but everybody knew about it.[25]

He came to regret his youthful scorn for what he recalled as the conservative academic psychology he was taught. There was "a gold mine in that academic thinking that, for all its sterility, could be mined in connection with psychoanalytic thought," he told Nancy and me. He had been too rebellious and too ready to write off what he had been taught. He hoped we would not make the same mistake. "Even from the viewpoint of learning the convictions and biases that need to be torn down by creative thinking," he told us, "one needs to pay attention to whatever preponderant systems are being presented. One should not be cynical by preference, by emotional slant, but only by real knowledge."

In a tribute to Reik on his sixty-fifth birthday, my father evoked his awe and anxiety on meeting Reik for the first time (in November 1938), placing himself in a genealogy of psychoanalysts going back to the master himself. As Reik had described ascending the steps of No. 19 Bergasse "with pounding heart" to stand face-to-face with Freud, my father, too, "carried a pounding heart up some steps, through a door, and into a small, simple study." There he first shook hands with Reik and began the treatment and training that changed the course of his life. Theodor Reik was the ideal analyst for my father, not only on account of his literary background and closeness to Freud but also because of his warmth and wit and his iconoclastic, provocative nature. These two independent souls came to admire, respect, and enjoy each other, both claiming loyalty to Freud while venturing in somewhat independent directions.[26]

Reik was also an accessible writer who, by the time he arrived in New York, had published several books in German and numerous journal articles. His most popular book, *Listening with the Third Ear*, published in 1948, best conveys his ideas about psychoanalysts as "psychological deep-sea divers." From Reik my father learned that it was "with the antennae of our own unconscious that we feel what is the essence of the thoughts and emotions of our patients, not with the tools of reasoning and logic," and that "what is heard with a third ear must be heard again and examined in the control room of reason. Both inspiration and perspiration have to be at work." Not for Reik—or for my father—the rigidly silent, withholding stance of many of Freud's followers, which had led Freud himself to declare that he was no "Freudian." In recounting to Nancy and me the stories he had heard from Reik, my father suggested that his own analysis was more a lively interaction than a monologue interrupted by occasional bursts of interpretation from behind the couch. One of the explicit lessons he learned from Reik about therapeutic technique was the use of stories—of which Reik had an inexhaustible supply—to convey an interpretation that might arouse resistance if conveyed more directly. As Dad put it, the listener may be the true subject, but through a story she is impressed into the role of observer. And she may often grasp the point more readily, with less defensiveness, because it is first made about others.[27]

One of my father's early papers described his experience of Reik's therapeutic storytelling. Dad had been postponing having children for several years because of concerns over responsibilities to his aging parents. Hearing this, Reik told him the following story. A mother bird and her three baby chicks began their winter migration from Europe south toward the Mediterranean and the warmer lands of Africa. When they got to the shores of the sea, the mother bird placed the first chick on her back and started the flight out over the water. Far from the shore, she asked, "When I am old and feeble, will you carry me on your back as

I am carrying you?" "Yes, Mother," the chick replied. Whereupon the mama bird dumped her dutiful chick into the sea. Returning to shore, she put the second one on her back and began the flight south once more. After a time she asked the same question and received the same answer. Whereupon she dropped this chick also to its fate. Returning for the third chick and setting out for the last time, she repeated the question. This chick replied, "Mother, I cannot promise you that. What I will promise is that when I have young ones of my own, I will carry them on my back as you carry me now." Whereupon the mother bird continued her flight with this chick south toward the distant shores of Africa.[28]

Evidently the tale did not convince Dad that he and Mom should have children right away because I was not born until two years after his analysis had ended—or rather stopped, because, as Dad said, analyses never end but turn from an interpersonal dialogue into an interior one—and we had left New York for Tennessee. But he absorbed the story's message, and, as Nancy and I grew older, he told us that story, repeated it to his patients, and even made a bas relief to illustrate its tough message. I do not know if he contemplated how his two adult daughters without children might interpret this tale. But in any case, he rehearsed it less often as he got older, and the bas relief eventually disappeared.

My father's training—or "didactic"—analysis, as he called it, was relatively short by twenty-first century standards: about two years, 350 hours, from November 1938 to September 1940, a common length in Europe and in New York at the time. Its purpose was not so much therapeutic as educational. It aimed to deepen his self-knowledge so as to enable him to respond in a productive manner to difficult issues of transference—those old feelings from past early relationships the analysand tended to project onto the analyst—and countertransference— the analyst's sometimes similarly inappropriate and anachronistic emotional response to the patient. After all, the interpretation of transference was at the heart of the psychoanalytic encounter as Freud had defined it; analysts were not free from its entanglements.

The training analysis also enabled him to experience personally what he would ask of his patients, an experience both Reik and Freud believed to be the most critical element in the education of an analyst, or a therapist of any stripe in my father's view. Starting in January 1940, after he had been in analysis for a little more than a year, he took four patients of his own into psychoanalytic therapy under Reik's supervision, along with three nonanalytic patients. And then in

September 1940, after his analysis ended, he joined Reik's seminar on psychoanalytic theory and technique, a small group Reik had gathered together for weekly discussions and case conferences.[29] Dad loved the idea that, for all their prudish hostility to Freud and psychoanalysis, his professors at Teachers College—no doubt Symonds was one of them—were secretly fascinated and on the sly approached him, their student, about gaining entrée.

Many years later, my father recalled one of his earliest patients whose symptoms had illuminated Freud's theory for him in an especially dramatic way. She was a deeply religious young woman whose leg had become paralyzed and painful so that she could not walk and ended up in the hospital. Physicians had been unable to find anything organic wrong with her leg. But during her treatment she revealed that her father had recently become senile and unmanageable and that she had placed him in a state hospital, an act for which she felt deeply guilty. Dad also learned that her father had been a shoemaker and had made shoes for her all her life. And that the pain and paralysis had begun shortly after she had hospitalized her father. At first Dad thought she must be joking because the symbolic connection between her shoemaker father, her sense of guilt, and a paralysis in her leg seemed too obvious. But she was not. "And what happened was that I became her analyst and within a few weeks she was very much better," he told me. He concluded that it was "strictly a transference cure, meaning in a sense that she had acquired a new father. This father was accepting her and not making her feel guilty and seemed to respect her. And that was enough to create a pseudo-cure. She didn't need the symptom [anymore]." Dad kept her on as his patient a while longer because she had many other problems. But she had no more pain, and she could walk.[30]

Through his experience with Reik, Dad fell in love with psychoanalysis, which he came to see as "the only general theory of normal and pathological behavior worthy of being considered a systematic and relatively complete treatment of human thought, feeling, and action."[31] Reik became a mentor, a father figure, a colleague, and even a friend whom he could help in practical ways, such as by going over with him the English translation of *Masochism in Modern Man*, a book my father admired. On Reik's behalf, Dad even carried on a brief correspondence in 1940 with the New York state attorney general and the state board of medical examiners in an unsuccessful effort to ascertain the legal status in the state of New York of the practice of psychoanalysis and psychotherapy by therapists trained by a medical practitioner but not possessing a medical

FIGURE 3.2 Theodor Reik in Estes Park with his granddaughter and Mom, circa 1948.

degree themselves.[32] Reik and my father stayed in touch for many years, and when we were living in Topeka, we visited Reik and his family in Estes Park, Colorado, on summer vacations.[33]

While my father was deeply influenced by his time with Reik and even shared some of his mentor's feelings of "outsiderness"—a feeling Dad at times embraced with pride rather than Reik's sense of grievance—he diverged from his mentor, too, especially as he came to focus on the theory and therapy of schizophrenia, an illness Reik rarely addressed.[34] He became more welcoming to the idea of psychoanalysis as a science, to science generally, and to efforts to test psychoanalytic theory and technique through empirical research, as well as to metapsychology, that discourse concerned with the structure of psychoanalytic theory (theorizing about theory). He was more respectful of the idea that psychoanalysis was possibly more useful as a research tool than as a mode of therapy. Just as Freud always considered that neurology and brain science would one day reveal the physiological bases of much psychoanalytic theory, Dad, too, came to believe that neuroscience and genetics would one day find the physiological correlates of *some* actions of the mind; he was interested in possible links between psychoanalysis and neuroscience, up to a point. "I'm all for the neurobiology of the brain," he wrote years later. "Neurobiology of the mind is a bit different. There is still something about thinking, feeling, acting by human beings which transcends chemistry. A relationship exists, of course, but I don't think neurobiology will tell us much about love and hate and inspiration. I don't think they'll ever build a computer with an artificial intelligence which will also include a human's consciousness or intentionality or feelings. The miracle of a computer that feels 'lonely' seems as far away as the edge of the universe."[35]

Although he would move on to mentors more theoretically inclined, my father remained deeply grateful for his time with Reik. The irony was that while the distinguished European lay analyst Theodor Reik never achieved acceptance within mainstream American psychoanalysis, my father, his student, who was also a lay analyst and far less learned, did achieve that acceptance. To Reik's credit, he did not hold it against Dad, even if he never let go of his anger toward the others.

My father remained loyal to Freud throughout his life, especially to the pessimistic Freud of *Civilization and Its Discontents*, a book written in the shadow of World War I and the Great Depression that left a deep imprint on his thinking. He never gave up his belief that humans, women as well as men, were essentially lustful, aggressive beings whose savage, biologically based instincts lay just below the surface of everyday civility. He did not call these instincts Eros and Thanatos, or life-and-death instincts, but he sometimes referred to libido and to mortido, a term he had learned from Bill Pious at Menninger. He was unmoved by the claims of neo-Freudians such as Erich Fromm and Karen Horney, who rejected Freud's nearly exclusive focus on intrapsychic conflict and accorded far more importance to societal norms and cultural values in shaping human subjectivity. For my father, the wide appeal of the neo-Freudians lay in people's longing for easy consolations and their preference for the inspirational over the instinctual. "To root significant aspects of man's thought and behavior in instinct, and more particularly in sex, contradicted more than a puritanical tradition," he wrote in a 1958 tribute to Reik. "It threatened to link together both the highest and the basest forms of human activity by tracing them to a common source in instinctual energy and instinctual aims. Is it any wonder then that the popularity of most neo-psychoanalytic movements stems from their efforts to attenuate the significance accorded to instincts and to man's biological nature?" Embracing Freud's comparison of the ego to a man on horseback "who has to hold in the superior strength of the horse" and who is sometimes obliged to follow the horse if he is not to be thrown, Dad emphasized the force of the unconscious and its power to guide human action. In his view, the neo-Freudian models were lame efforts to counter Freud's profound but humiliating truths. He was scornful of those he called "the culturalist theorizers and religious compromisers," as well as those ego psychologists who attributed "such autonomous powers of control to the ego as to relegate the id and all its unconscious instinctual drives to the position of the gentle, housebroken dog who only pretends now and then, when the milkman arrives, that it is a wolf."[36]

Still, Dad believed, following Freud, that the animal forces of human nature could be channeled into creative work and creative relationships as sources of deep pleasure and joy. If his view of life was essentially tragic, he would also increasingly embrace the utopian possibilities of science and art. He believed in the value of struggle—struggle against conformity and against adaptation to an unjust society, struggle to find one's own way of being in the world and to find others who shared one's values. In the face of the American ego psychologists' post–World War II emphasis on adaptation or adjustment to "reality" as a therapeutic goal, he would espouse a far more critical view, a legacy, I believe, of his time at Teachers College. Even the horror of juvenile delinquency, he wrote memorably in 1958—perhaps recalling "bad boy" Frank from his Teachers College days—often represented "a living pressure against social injustice and inequality, destructive and dangerous though it may be. But apathy, ignorance, and Miltown [blockbuster tranquilizer launched in 1955] may turn out to be greater dangers." That radical claim, in the 1950s, from an orthodox Freudian psychoanalyst, was considered noteworthy enough that of all the distinguished speakers at the 1958 tribute to Reik—including the political philosopher Herbert Marcuse, the literary critic Alfred Kazin, and the social psychologist Goodwin Watson—my father was the only one to be quoted the next day in the *New York Times*.[37]

He was not alone, of course, among those in the United States who called themselves psychoanalysts in the fifties and sixties to reject the ego psychologists' emphasis on social adaptation and conformity. Nor did an emphasis on the autonomous powers of the ego—a centerpiece of ego psychology—necessarily foreclose a more critical stance. But most historians of the subject argue, as Dagmar Herzog has put it, "that postwar American psychoanalysis was a normative and normalizing enterprise," an instrument of conformity and social control. Or in the words of Eli Zaretsky, "a veritable fount of homophobia, misogyny, and conservatism."[38] Mavericks such as Dad's acquaintance Robert Lindner, the author of *Prescription for Rebellion* (1953) and *Must You Conform?* (1956), a psychologist, psychoanalyst, and fellow Reik analysand whom my father admired, were definitely on the margins of the profession.

Dad mostly stayed out of the larger theoretical debates that roiled the psychoanalytic world in the postwar decades, except as they concerned the theory and therapy of schizophrenia and the training of psychologists and psychoanalysts. He would become less pessimistic as he grew older and even describe himself as "a constant optimist" when he was in his nineties. Though he never abandoned his sense of humanity as essentially barbaric, he nonetheless became more hopeful, more attuned to the intrinsic value of playfulness, joy, and art. He also turned more to the possibilities of science, not only to help remedy humanity's ills but also as a source of wonder and of awe. He learned a lot from the artists he came to know. Some of them considered him an artist, too.

Dad's comfort with change was tested between October 1941 and July 1946, when he and Mom—and then Nancy and I with them—moved four times in the space of five years. We lived the kind of gypsy life my father had grown up with when he had changed houses every two years. By October 1941, he had finished the coursework for his PhD, and all that remained was his doctoral dissertation, a project he was not eager to begin. Offered a position as assistant professor of clinical psychology at the University of Tennessee in Knoxville and a chance to build a university psychology clinic as its director, he decided to accept. Although he was seeing patients and still working as a research assistant for Percival Symonds in the adolescent fantasy project that could have counted for his dissertation, he continued to find his association with Symonds onerous. The position at the University of Tennessee was a challenge that allowed him a graceful escape. He discovered that he liked teaching well enough, was a popular professor, and that there was much demand for services at the psych clinic he helped to create.

Two months after he and Mom arrived in Knoxville, however, the United States entered the war. News from the battlefields of Europe and Asia and terrifying reports of the escalating persecution of Jews and other minorities throughout Europe cast a shadow of fear and uncertainty over the nation. Into this anxious atmosphere I was born on May 31, 1942, at the Knickerbocker Hospital in New York City, where Mom had returned to give birth in the company of her mother and her best friend, Aline. Dad learned about my entrance into the world through a telegram delivered to him while he was on a train en route to join Mom. She had remained deeply attached to her mother, Sabina, who had raised her more or less as a single mother once Abraham began to succumb to Huntington's chorea. Dad, however, thought Mom too attached and later blamed Sabina for overpowering her and keeping her from becoming independent, although Mom's insistence on giving birth in New York over his protests suggests otherwise. In any case, Dad later recalled the "transforming experience" of becoming a father.

"It was like falling in love plus growing up plus fighting off all the dangers of a somewhat unsettled world," he wrote. He evidently took to being a father at once, and I never heard him express regrets or reservations of any kind.[39]

In April 1943, Dad was offered the position of chief of a test and research unit for the Army Corps of Engineers, with the task of developing psychological testing programs for civilian personnel and offering counseling to employees. They were having personnel problems and recruited Dad to help set up programs to assist employees and develop research. We left Knoxville for Washington, DC, in April 1943 and soon settled into a new Parkfairfax complex, then under construction across the Potomac in Arlington, Virginia. Dad disliked his boss and was only mildly interested in the work but put in considerable effort. "That's kind of a characteristic, I guess, of mine," he wrote later, "that I can get involved in almost anything. It can be dull, it can be business, it can be a remote kind of experimental psychology, and I can still get fascinated by it. To me it's all like building those old radio circuits. You just tinker around to see if it works."[40]

And then, restless and eager to get into uniform and contribute more directly to the war effort, Dad got himself appointed a lieutenant commander in the Research Division of the Bureau of Naval Personnel, with a job doing research on psychiatric screening procedures to identify "maladjusted personnel unfit for general duties or specific programs such as submarine or amphibious duty."[41] (Because the Navy lacked enough psychiatrists to interview recruits personally, good screening tests were a high priority.) As the only psychoanalyst in the Bureau, he once again enjoyed a certain distinction. Since they did not know how to use his skills, they gave him leeway to develop his own programs. He decided to devise a research project that he could use for his PhD dissertation. He invented a test that was supposed to discriminate between those who were vulnerable to psychological breakdowns and those who were more resilient. Dad always dismissed this project as a pedestrian exercise, useful only in qualifying him for his degree. But even in this study, as in "The Story of Frank," there are hints of his critical and questioning style of thought. For instance, he noted that while the existing tests aimed to identify psychological correlates of success in the Navy, they sometimes wrongly labeled as maladjusted individuals who were in fact "quite adequate." As he put it, "Summation of symptoms present may represent but a fragment of the total picture. What of the assets?" He recommended changes in research methodology that might better capture "actual success in adjustment rather than psychiatric prediction of success." Valuable or not, this piece of original research, published as part of the Navy's personnel research and test development program, lifted him past the last hurdle of his graduate studies. In 1948, at the age of forty, ten years after he had entered Teachers College, he finally received his PhD and became a doctor of philosophy once and for all.[42]

FIGURE 3.3 Milton served in both the U.S. Army Corps of Engineers and in the U.S. Naval Reserve; circa 1944.

FIGURE 3.4   Henry served with the U.S. Army Air Forces; circa 1944.

And it was while we were living in Parkfairfax that Nancy was born, on July 19, 1945, three weeks before the United States dropped its first atomic bomb on Hiroshima, Japan. Once my parents persuaded me that I would not be forgotten with the entry of this intruder into my former paradise where I had been the center of the universe, I decided to stop my efforts to poison her or otherwise contribute to her demise. I became her protector and defender, and she became my playmate, my comrade in arms, and my best friend.

Once again a fortunate encounter shaped my father's trajectory. Or rather, he increasingly narrated his life in ways that framed certain events as strokes of luck. With the war over in August 1945, he was trying to finish his dissertation and decide what to do next when he met David Rapaport, who was to have a profound impact on his life. Rapaport was a Hungarian-born wunderkind, an expert on diagnostic testing and a synthesizer and systematizer of psychoanalytic theory, especially of ego psychology. Hired at Menninger in 1940, he created an influential research group—and ultimately a research department—within the Menninger School of Clinical Psychology, which he also founded. Rapaport, like Dad, was a clinical psychologist who found a home within a psychoanalytic setting. He helped make testing a part of clinical practice and also helped to establish clinical psychology—working with patients—as an important field within the discipline of psychology and, at Menninger, an equal partner with psychiatry. His leadership of the Research Department while he was at Menninger and his encouragement of collaborative practice and of each of its members left a lasting impact on those who worked with him.

As Dad recounted many years later, Mike Dunn, who was already at Menninger, arranged for him to pick up Rapaport at the Washington DC train station and drive him to the hotel where he was staying while attending a conference. Dad was immediately smitten, finding Rapaport brilliant, stimulating, and enormously well read in psychoanalysis, experimental psychology, and philosophy. Although Rapaport was a few years younger than my father, he had much more experience and knowledge. Dad felt that here was a man he could truly respect and admire. He felt that Rapaport liked him, too.[43]

In my father's narration, meeting Rapaport in Washington DC was the fortuitous event that led him to Menninger. But I cannot help feeling that as both Mike Dunn and his brother Henry and his family were already in Topeka— Henry had been given a 90-day training program in psychiatry in the Army Air Forces during the war and wanted to pursue a full psychiatric residency afterward—Dad might well have joined them even if, thanks to the timely intervention of Dunn, he had not picked up Rapaport at the Washington DC train station and driven him to his hotel in the spring of 1946. Still, it was after that meeting that my father, in May, boarded a train to Topeka for interviews with Karl and Will Menninger, among others. And during the visit, Karl, who was then the chief of Winter, offered Dad a position at the hospital. As the acting director of psychological services, he would see patients, supervise residents, teach, and possibly engage in research. Two months later, we were on our way: one-year-old Nancy, four-year-old Alice, Mom, who was thirty-two, and Milton, leaving the East Coast for the great American heartland. At the age of thirty-eight, my father was about to begin an era of his life in which he finally felt surrounded by people he esteemed. Here in the middle of conservative, provincial, Protestant Kansas, my Jewish, Brooklyn-raised, dyed-in-the-wool New Yorker father was about to become truly himself.

4

The Slap, Explained

Psychoanalytic history has always been embodied in relationship, and these pairs have taught us what little we know about how therapy works.

—Gail Hornstein, *To Redeem One Person Is to Redeem the World*

Standing on the sidewalk in front of the old Menninger Clinic in Topeka that snowy November day of 2014, I found it hard to imagine the charisma that once surrounded the place. The pale green American Foursquare house before me looked forlorn in the wintry sunlight. Any image of Topeka as the "Psychiatric Capital of the World," as a highway billboard had once proclaimed, seemed remote.[1] Yet in my father's memory this clinic and all the other Menninger institutions associated with it remained the lodestar, the place he finally wanted to be. At the moment we arrived in the summer of 1946, Menninger was becoming a community brimming with intellectual excitement and a sense of being part of a great historic experiment in improving the care and understanding of mental ills. At the core of this firmament were the Menninger brothers, Karl, the mercurial visionary, and Will, the organizational genius (and former chief of army psychiatry); the elegant Margaret Brenman-Gibson, one of the few women analysts on the staff and a training analyst, too—a critical position since training analysts were the gatekeepers into the profession; erudite Merton Gill; the gentlemanly aristocrat Robert Knight, greatly respected as a clinician and researcher; the enthusiast Arthur Marshall, who had managed army

hospitalizations in Europe during the war and was now playing a major administrative role at Menninger; and the scholarly Sibylle Escalona. And of course the
inimitable firebrand David Rapaport.

Topeka became a haven, at least for my father, who was instantly caught up
in the excitement and enthusiasm of Menninger as it began its great post–World
War II expansion. As soon as we arrived he was embraced by the Menninger
community—or family, as the Menninger brothers liked to describe it—with
brilliant colleagues he admired, intellectual challenges, and respect from the
younger, mostly male students and psychiatric residents, as well as from Karl and
Will Menninger. In many ways our years in Topeka were idyllic, not only for Dad
but also for Nancy and me, and perhaps even for Mom. Rather than fleeing to
the suburbs and a more isolated life, as did many of the white veterans returning
from the battlefields of World War II, we became part of close-knit social world
filled with optimism and a sense of mission. Members felt they were pioneers at
the forefront of a new psychiatry and psychology, with the will and the means
to transform the care of the mentally ill. Despite the fact that we were Jewish in
largely Protestant Kansas, there were enough Jews at Menninger to make the community an oasis, enough so that the enterprising Arthur Marshall began ordering
bagels and lox to be delivered to his home for communal brunches every Sunday.

Topeka in the late 1940s was a city of approximately eighty thousand people, of whom about 92 percent were white, racially segregated in most respects,
though public buses were not, and the local NAACP had fought segregation for
years, in cases that culminated in *Brown v. Board of Education of Topeka*. Some
at Menninger were involved in civil rights actions too, including the "Winter
Wives"—women married to male staff and faculty at Menninger—who successfully challenged the restriction of Black moviegoers to the balconies. But a
hierarchy of race and gender shaped life at Menninger, in ways I was too young to
recognize at the time. Most of the analysts and other psychiatrists, residents, and
students were white, while many of the service workers were Black. The former
were also mostly male. At one point Dr. Karl became so worried that the male
staff and residents were neglecting their families because of their total immersion
in work that he took it upon himself to send out a memo reminding husbands to
pay attention to their wives.

Nancy and I did not feel neglected. And without brothers, we did not experience the gender inequities that some of our friends confronted early on. We
did not compete with brothers for attention from our parents. We roughhoused
with neighbor boys in our backyards. We thought we were just as smart as they
were. If there was anti-Semitism in Protestant Topeka, we Jewish kids in the
oasis of Menninger were mostly unaware of it, just as we did not question why

FIGURE 4.1 Some Wexlers in Topeka. All from left. Back row: Milton, Nathan, and Henry; middle row: Liza (Henry's wife), Sabina (Sabin), Mollie, and Leonore; front row: Willa (Henry and Liza's daughter), Nancy, and Alice, circa 1947.

all the kids at Randolph Elementary were white. We learned much later that Karl Menninger's progressivism on many fronts stopped short of supporting Linda Brown in the *Brown v. Board of Education* case. Nor was he willing to resist anti-Communist Cold War pressures from Washington as some others at Menninger were.[2] The complexities of Kansas history would come as a surprise long after we had left the state.

In contrast to the majority of young people who flocked to Menninger at the close of World War II to do a residency in psychiatry, get psychoanalytic training, or pursue a PhD in clinical psychology or social work, my father came as a member of the staff. He was older than most of the residents and had completed his training analysis. He was not a student but a teacher, accorded the status of professor at the University of Kansas. He was an object of admiration to many of the residents—a "clinical conquistador," as one called him. "He was highly regarded," his friend and colleague in Los Angeles Mike Leavitt later told me.

"He had a certain aura about him; he was impressive to people. He was considered a kind of original mind."[3] Herbert Schlesinger, who later became an eminent psychologist, recalled him as "tall, good-looking, with unimpeachable manliness," noting that "we were all in awe of him because he was what most of us wanted to be."[4] Gerald Aronson, an analyst who also became a close friend in Los Angeles, recalled my father as patient and wise, someone "who made me feel that I knew something." Gerry recounted to me how my father would let him "unspool the thread, never intrude. And then he would add a coda to highlight a different perspective. He was never critical," Gerry told me. "He would say, 'Well, have you ever thought about it in this way?'"[5]

While the residents and students looked up to him, Dad's relationships with other staff members were more complex. "There was kind of an inside group and an outside group, and I was somewhere in between," he recalled. "I didn't get into that very special inside group that consisted of Rapaport and Brenman and [Merton] Gill and people of that sort, but I was always there and always acceptable," he wrote later. "I even had a certain kind of peer relationship with them but not that kind of tremendous intimacy that existed in the group."[6]

When my father presented his paper on Nedda before the Topeka Psychoanalytic Society in June 1950, he did so before a society that, despite being located in Kansas, was second in prestige only to those of New York and Chicago. From his opening salvo he portrayed himself as someone to be reckoned with, challenging the ideas of Frieda Fromm-Reichmann and other prominent therapists, several of them women, who had worked with patients with schizophrenia far longer than he had[7]:

> Love in all its myriad forms still stands as the principal prescription for schizophrenia. Affection and sympathy, tenderness and approval; these are the medicines of choice. Dosage, of course, depends only on the capacity of the therapist to give, and he is the best therapist who has the greatest libidinal resources. To all of which one might ask: How can it be that such a commodity so lavishly expended can be so potent? Or, how can it be that so potent a commodity is so lavishly and incautiously prescribed? . . . Few things are so calculated to terrorize at least some schizophrenics as untimely affection, however real and unambivalent. Indeed, the same is true of some skittish children who can be won only by disinterest. If the veriest show of encouragement can be translated by a neurotic in the analytic situation as a seduction, its counterpart with the

schizophrenic may be perceived as a rape. Love is no doubt an effective medicine but, as Freud indicated, many effective medicines are poisonous in nature. We must have caution how we use it.[8]

It is unclear to me how much Dad's Menninger colleagues, apart from Bill Pious and the resident observer George Harrington, knew about the Schizophrenia Project or about his unconventional treatment of Nedda prior to his presentation to the Society. Their response suggests that they had not known much, for his paper elicited not only admiration but also shock at learning that he had slapped her, "sometimes very hard." He acknowledged that his narrative risked being interpreted as "an endorsement of acting out on the part of the therapist," but the question *he* wanted to emphasize was why his patient had gotten better.

Unfortunately for Dad, "the slap" became the focus of discussion. Dr. Karl worried that publicly acknowledging the use of physical violence against a patient without explicitly repudiating the practice would give the impression that Menninger endorsed it; he especially wanted to avoid this misconception in light of what he said was the widespread approval of such behavior by American parents with respect to their children. He was astonished, he told my father privately, by "the triumphant note of victory and successful accomplishment" in letters he received from parents who were proud of the harsh methods they used on their children. He was adamant that my father either disavow the slap or omit it from the paper altogether.[9] Taking a different tack, the visiting Norwegian psychoanalyst Trygve Braatøy, whom we would later visit in Oslo, argued that "the slap" indicated the tremendous strain of treating psychotic patients such as Nedda, implying that my father had temporarily lost control of himself. For that reason alone, Braatøy insisted, it must be included in the paper.[10]

In the published paper, "The Structural Problem in Schizophrenia: Therapeutic Implications," my father did acknowledge "meeting force with force" in response to Nedda's assaults. But his emphasis was on the impact of his prior shift in therapeutic strategy, from trying to dispel her harsh self-accusations to agreeing with them. He felt this was the major turning point in her therapy. He theorized that when he made this turn toward echoing her self-criticism, Nedda became calmer and more rational because he was speaking her language; she felt understood. He had succeeded in getting her to listen to him. He speculated also that he had helped her feel less frightened of her own strong impulses since *he* was in control. Gradually he made it possible for her to develop distance from her self-accusations, allowing her to reflect on the extremism and irrationality of the inner voices that told her she was bad. Her assaults gradually diminished.[11]

Nedda's improvement became the starting point for his theory of schizophrenia as an illness marked by the internalization of a harsh and punitive superego ("sex is bad," "sexual thoughts are sinful," "marrying outside the church is bad") engaged in unremitting warfare against instinctual impulses (the id) on a battleground in which the ego was more or less squashed and defeated. In other words, it was Nedda's savage, self-punishing superego that was at the root of her illness. As he put it, that superego was "a primitive, archaic structure in which the primal identification (incorporated) figure of mother holds forth only the promise of condemnation, abandonment, and consequent death." The therapist's task was to break through to and resurrect "remnants of a reasonable ego."[12]

Many years later in his memoir, my father offered a more dramatic version of Nedda's breakthrough to rationality, reversing the chronology he had laid out in both his presentation to the Topeka Society the summer of 1950 and in the published paper. The memoir version was the story he had told Nancy and me. "Nedda was sitting in a low, cushioned chair and I walked over to offer her a pad on which we scribbled notes to each other," he wrote. "Suddenly, as I faced her, she kicked me in the testicles with such force that I doubled over in pain and, without a second's thought, I slapped her face." Not hard but enough to grab her attention. "Why'd you do that?" she had asked him. "So far as I knew, it was the first clear, intelligible sentence out of her mouth during her many years of hospitalization, certainly the first I had heard from her that made sense." He subsequently theorized, as he had earlier, that the slap had reinforced her superego and told her that "there was a force out there that wouldn't put up with her violence or her sexuality." By allying himself with her self-punishing superego, he had allowed "the rational ego to peek out and say an intelligible sentence, knowing that the frightening forces of the id would be contained." It was only then, in this version, that he decided to shift toward a stricter, harsher disciplinary stance. In this later account, the slap did not follow his change in therapeutic strategy as a deliberate extension of his tough verbal discipline, as his paper implied; it was an accidental response that inspired the shift.

Which account is closer to the historical events? Surely the monthly reports written at the time are the more reliable. Yet the version he had described to Nancy and me and the narrative written much later have a concreteness that the earlier version lacked. I feel certain that Nedda did kick my father in the testicles at some point but probably *after* he had shifted his strategy, as the monthly reports suggest, and that he inadvertently dramatized in memory what had been a more gradual change. We may never know, and unfortunately Nedda, the central figure in the drama, has left no clue.[13]

౷

In placing the blame for Nedda's psychosis on her harsh, unrelenting superego—a superego that he claimed represented an internalized image of a rejecting, punitive mother—my father drew on the thinking of John Rosen and, even more, of Bill Pious, who shaped his ideas from early in the project and for many years after. Bill and Dad met together weekly, usually on Saturday night over lox and bagels, discussing the case and theorizing about schizophrenia, leaving Mom and Jess Pious, Bill's wife, to entertain themselves while a babysitter kept Nancy and me occupied at home. A native of Connecticut with a medical degree, Bill had come with Jess to Topeka to train as a psychoanalyst in 1941 and had begun teaching in the Menninger School of Psychiatry in 1946. He also became a highly respected training analyst, but he was known as a man who did not suffer fools gladly. In fact Bill was shy, as Dad was quick to realize. Dad was one of the few at Menninger to socialize with Bill, who preferred fishing and rock collecting to parties.[14]

FIGURE 4.2 William Pious, a highly respected training analyst at Menninger in the 1940s, circa 1955.

Pious described the pathogenic process in schizophrenia as "a sudden vacuum in the patient's psychic apparatus." This vacuum or "emptying," according to Pious, was the consequence of a defective superego that could not contain in a normal way the aggressive and destructive energies of the id—energies that Pious (after the analyst Paul Federn) termed "mortido" and that was more or less equivalent to Freud's death instinct. This mortido invaded and overwhelmed the entire psyche, leading to sensations of dying or being dead. To combat those sensations, all the life energies, or libido, would withdraw from attachments in the external world to try to combat the mortido flood.[15]

My father did not entirely follow Pious's rather turgid formulation, but he did not find their differences irreconcilable. They both agreed that the problem was basically a defective superego. So whether one conceptualized a superego that was too aggressive toward the ego or one that was too weak to contain the mortido and protect the ego, the devastating consequences were the same.

He also put forth a second idea, suggested by one of Pious's articulate patients. I'll call this patient Mr. M. Pious explained Mr. M.'s idea as follows: when functioning relatively well and communicating with Pious, Mr. M. experienced what he called the "mental image of the analyst" clearly. When he was not functioning well, the mental image became distorted. Pious concluded that the image functioned for Mr. M. as a kind of superego. Pious believed "the development of the 'image' and of the capacity to retain it were decisive in sealing off the pathogenic process." From the time that the image became firmly established, Mr. M. had been able to work with Pious without falling into a psychotic episode.[16]

Building on this formulation, my father suggested that, by agreeing with Nedda's severe moralizing, he had enabled her to begin holding on to a clear mental image of him. By agreeing with her strictures, he had made himself into a kind of substitute or external superego that helped control the feelings she experienced as overwhelming and dangerous. He had thereby enabled her to get some distance from those feelings, to begin thinking more realistically about them and even to question her own extreme claims. Or so he argued at the time.

It is not surprising that Pious, Rosen, and my father conjured up a pathogenic mother as the villain in the etiology of schizophrenia—or at least "a pathological relation of the mother to the patient in his first months of life." The term *schizophrenogenic mother*, coined almost parenthetically three years earlier by Fromm-Reichmann, did not, as of 1950, have much play.[17] But blaming the mother had considerable purchase in psychoanalytic thought generally, as well as in U.S. popular culture, especially during the Cold War years. As the analyst Silvano Arieti wrote in his 1974 textbook *Interpretation of Schizophrenia*, "In the writings of a large number of authors she was described as a malevolent crea- ture and was portrayed in an intensely negative, judgmental way."[18] In alluding to the unconsciously hostile mother, my father repeated a common psychoanalytic shibboleth about the past history of patients with schizophrenia, but ultimately he was not focused on the past; he was far more interested in what was happen- ing with his patient in the present.

Reading my father's unpublished and published material about Nedda, I notice that whereas his monthly reports during the Schizophrenia Project accorded great therapeutic importance to Nedda's expanded social activities and companionship outside the hospital, his oral presentation to the Topeka Society and his published papers all but ignored them. I cannot help feeling that the prestige accorded to metapsychology and psychoanalytic theorizing at Menninger in the early 1950s made it tempting for up-and-coming analysts there to highlight such theory above all else. The social support that his monthly reports emphasized as highly therapeutic went mostly unremarked.

Soon after his June presentation to the Topeka Society, my father began work- ing on his follow-up paper for the December conference on the psychotherapy of schizophrenia at Yale. Considering how recently he had entered the world of psychoanalysis and his limited experience treating schizophrenia, he might have found his fellow presenters daunting; all had medical degrees and several had uni- versity affiliations as well. But if he did, he left no indication of it. He had met Fromm-Reichmann the year before in Topeka, when she had come to speak at Menninger about her forthcoming book, *Principles of Intensive Psychotherapy*, and he was a discussant on her presentation. Frieda Fromm-Reichmann by this time was a greatly respected and admired figure within the psychoanalytic world. Since emigrating from Germany in 1935, she had made Chestnut Lodge a sought after private psychiatric treatment center and like my father, she was one of the relatively small community of analysts and psychiatrists who believed in treating schizo- phrenia through psychotherapy. *Principles of Intensive Psychotherapy* elaborates in her characteristic straightforward, non-jargony language her theoretical and tech- nical approaches to treatment.

As a member of the school of interpersonal psychoanalysis founded by Harry Stack Sullivan and recognized more as a clinician than a theorist, Fromm- Reichmann must have expected some disagreement when she came to Menninger, with its reigning blend of orthodox Freudianism, ego psychology, and Midwestern pragmatism. However my father's criticism was so severe, according to his memory years later, that he drove her to tears and felt compelled to apologize. Indeed he approached her book less as a colleague seeking new insights than as a debater arguing the other side, starting by announcing that the book "both inspires and disappoints" and noting soon after that her logic was "faulty and the level of explanation or definition inadequate." After nine pages of criticism, in which he attacked both her and Sullivan's interpersonal theory, he concluded his discussion with a patronizing nod to "her beautiful and succinct case materials" and her courage in keeping alive a hopeful attitude toward the psychotherapy of schizophrenia.[19]

The gracious Fromm-Reichmann apparently brushed off his criticism, especially after he apologized, as the arrogance of an ambitious young analyst eager to show off his critical chops before his Menninger colleagues. Three years after Yale and the Midwinter Meetings of the American Psychoanalytic Association, where they had again shared the podium, she would be on yet another panel with him, acknowledging in a warm letter that "we are all very glad that you are one of the discussants of the papers" and that she looked forward to seeing him.[20]

Like his earlier paper, my father's Yale presentation, "The Structural Problem in Schizophrenia: The Role of the Internal Object," rehearsed his two-year treatment with Nedda and once again argued that, in the face of a patient's dread of his or her own overwhelming impulses, a strong hand of prohibition from the therapist could be reassuring. "The exhausted, panic-stricken soldier in the holocaust of battle wants none of our tinned meats and canned music," he wrote. "More likely he prays for an all-powerful God or a tough top-sergeant who knows up from down and says so out loud." To those who took umbrage at this idea as a return to medieval treatment methods such as the whipping post and the dungeon, he urged that they might reconsider "the scientific objectivity of some of our own procedures." Besides, the superego represented not only the conscience but also "the lifetime accumulation of ego ideals" and the protecting powers of the parents. For the therapist, siding with the superego meant identifying not only with prohibitions but also with ideals, hopes, and dreams.[21] This time he also took issue with the prevailing psychoanalytic notion that schizophrenia represented what Freud called a regression to primary narcissism (that is, a return to an earlier state of total self-absorption and lack of differentiation between the self and nonself), with a loss of both internal and external objects or images.

FIGURE 4.3   Milton at the "Conference on Psychotherapy with Schizophrenic Patients" held at the Department of Psychiatry, Yale University School of Medicine, December 6, 1950.

Photograph by Lawrence Kubie.

In his view any visit to a mental hospital gave quite the opposite impression. He argued that most people who had schizophrenia were deeply engaged with their mental images, which represented "internalized parental figures dressed in the variable garb of one's childhood mythology." As my father put it poetically, "The ghosts of the past are certainly not laid to rest nor generally immersed in some retreat into an objectless, oceanic narcissism." However, "the very special quality of these internal objects in the schizophrenic is their enormously punitive and unloving relationship to the ego." As he had claimed in his earlier paper, these were "paradigms of the cold, unloving, hostile mother." The one incontestable model that therapists could offer to replace this punitive internal object was themselves. "Until, as therapists, we have succeeded in behaving in such a way as to make it possible for the patient to build within herself a clear and steady image of another kind of person," he concluded, "no consistently effective ego functioning is possible."[22]

This time Dad showed a draft of his paper to Dr. Karl before he presented it in New Haven, receiving in return another terse memo, this time questioning why he gave so much credit to Bill Pious and so little to others at Menninger for shaping his thinking. When his paper appeared in print, in the *Bulletin of the Menninger Clinic* (November 1951) and a year later in the volume of papers and commentaries from the conference, it included a fulsome encomium to the staff of the Menninger Foundation and the Winter V.A. Hospital, with a quote (in the volume) from Karl's *The Human Mind* making the point that schizophrenia was not a hopeless disease, that "a considerable number" of patients do get well with prompt and skillful treatment. Dad also credited the Menninger atmosphere of optimism toward the treatment of schizophrenia, although in his heart, I believe, he mostly thanked Bill.[23]

My father's Yale paper elicited strong disagreement from the two discussants, Robert Bak, a psychoanalyst from Budapest who had treated the famous Hungarian poet Attila József for schizophrenia, and Ludwig Eidelberg, an analyst also associated with the New York Psychoanalytic Institute. This time the criticisms centered not on his admission of meeting force with force, as in Topeka, but on his entire conception of the schizophrenic process and the therapeutic strategy he had employed with Nedda, though they did not question that she had made a remarkable improvement up to that point. In Bak's view, the major fault line in schizophrenia was an ego disturbance, and the superego, "if we can speak of it at all," played a subordinate role, a view shared by most of the panelists. Both Bak and Eidelberg questioned what they called my father's "authoritarian attitude" toward his patient and what he claimed was the effective element in the therapy, namely his temporary identification with her harsh superego as a means of enabling her to develop a firm mental image of her therapist and thereby distance herself from her delusional views.[24] But whatever the criticisms, some of which he would come to agree with, my father had shown— and Bak and Eidelberg did not question—that intensive, daily, one-on-one psychotherapy, talk therapy, over a period of a little more than two years had enabled one long-term hospitalized patient with schizophrenia, previously unresponsive to any kind of treatment, to leave an institution and live reasonably well in a supportive family environment for a period of months if not yet years.

Nedda returned to Topeka in April 1951 with her husband and again met with my father, who found her functioning "at a fairly effective level." Her husband reported that she was capable of maintaining the household and fulfilling the duties of a housewife—banking, shopping, cooking, cleaning—with only occasional difficulties "due to her failure to understand the environment in which she lives." She told my father that in times of stress, she would fall back on some letter, picture, or object that he had given her as a means of reassuring herself and of bringing into focus the world around her. It is unclear whether Nedda's husband or my father was the one who neglected any mention of her emotional well-being and whether her anxieties and sense of herself as evil had subsided.[25] Not surprisingly for 1951, her ability to perform as a conventional housewife became a measure of her psychological well-being, whether she enjoyed or desired those domestic duties or not.

Fourteen years later, in a discussion of the office treatment of patients with schizophrenia, my father noted the contrast between Nedda's clinical improvement and the persistence of psychotic thought patterns as shown in the results of subsequent Rorschach tests. "For long years after this recovery," he wrote, "the Rorschach test showed no change whatsoever . . . even with the most amazing clinical changes." He concluded that the Rorschach was not as sensitive a test as many clinicians made it out to be.[26]

My father never forgot Nedda, and Karl Menninger never forgot her case either, or his own version of it. Writing nearly four decades later, Dr. Karl reminded

Dad that he had made some wonderful contributions and that "long, long ago you took a case at the VA Hospital for psychological treatment whom the rest of us didn't think could be treated that way or would likely ever get well. But Milton had hope, the most important element in the medical armamentarium." Dr. Karl was unsure whether Nedda had recovered. "Whether she did nor not, she should have and I have had him [Milton] in my mind ever since as one of those doctors that behaved like a doctor and went ahead bravely and persistently and successfully on a lonely journey."[27] Dad was proud of that letter and showed it to friends, as if to remind himself that his struggle against another "incurable" disease, Huntington's, was also worthwhile.

Nedda remained a touchstone of my father's life and, in some ways, of my sister's and mine: she was the first of our father's patients whom we met and one whom we heard about all our lives. How long she was able to function without extreme disturbance and the quality of the life she lived after she left Menninger remain unclear. My father regretted that she had had to leave Topeka prematurely and return to a home environment lacking the social and psychological support she needed to keep her demons in check. His unpublished monthly reports, filed away in the Menninger archives, suggest that the social support structure he had created for her in Topeka had been critical to her improved functioning, though his publications mostly credited therapeutic strategies such as establishing strict boundaries, identifying with her punitive superego, and enabling her to internalize "the image of the therapist."

For many years she wrote letters to my father, long, rambling, repetitive letters with many irrelevancies. She told my father that when she had returned home from Topeka, she and her husband had finally been married in the Catholic church, something she had long desired. But then nine years later came "the wreck" which she did not explain except to indicate that it destroyed the marriage. Eventually the letters started coming from a board-and-care home in a different state. She had a guardian during the last twenty-four years of her long life. When she was approaching ninety, she expressed a continuing sense of loss despite the many decades since her divorce. My father kept Nedda's letters, and after he died I discovered them among his papers. To ensure that they would not get lost or damaged, I placed them in a storage facility in Venice, California. In November 2014 a fire raged through Extra Space Storage and the letters were destroyed. Only a few preserved in the Menninger Archives and one or two I stashed in a filing cabinet in my Santa Monica apartment have survived.

5

Ex-Topekan

Thinking is never a purely "cold" process.

—David Rapaport, "Toward a Theory of Thinking"

S pring 1951. My father came into the bedroom that Nancy and I shared in Topeka and sat on the edge of my bed. He asked us how we would like to go on a big ship across the Atlantic Ocean. He told us about a man with schizophrenia, like Nedda, who lived in Norway and was going to be his patient for the summer. We were all going to Norway together, and we could have a vacation there, too. Some time later he told us that we would not be coming back to Topeka, except for a brief visit, because we were moving to Los Angeles. He said he had always dreamed of returning to California. Now we were going to live there for good. He said we would have a great adventure and promised that California would be even nicer than Kansas, although I doubted that could be true.

On January 30, 1951, my father wrote to David Rapaport that he was leaving Topeka "because recent and urgent financial obligations make it impossible to stay." He hinted at ominous events to come. "I do not know if I can solve these problems elsewhere but I should like to try," he told Rapaport. "California is presently my choice, partly because I like the West, partly because personal (family)

involvements make the East very difficult, if not impossible, to consider."[1] There
is no other indication in his letters of the news that he and Mom had received in
late September or early October 1950 from New York. In the Park Avenue office
of a Dr. Bender, a neurologist, at separate appointments, all three of Mom's older
brothers—Jesse who was forty-six, forty-four-year-old Paul, and Seymour, at
forty-two—were diagnosed with "a neurologic disorder which is characterized
by chorea, mental changes, familial and probably hereditary factors." The neurologist neither knew the etiology nor did he give the condition a name, "but it
is believed to be a heredo-familial process in which there is a slow but progressive
degeneration of the brain."[2] Mom knew exactly what it was. Her brothers had
the same disease their father Abraham had died of in 1929, the name of which
was marked on his death certificate at the Central Islip State Hospital on Long
Island: Huntington's chorea.

Dad consulted Leon Bernstein, our neurosurgeon next-door neighbor, who
told him what he knew. But he spoke to few others in Topeka. He knew that
general knowledge of his wife's brothers having this severe hereditary illness
could stigmatize the entire family and prompt unwelcome questions from his
colleagues. They might start scrutinizing Mom for symptoms and question why
Mom and Dad had had children if they knew of her hereditary "taint." Public
gossip would certainly have had repercussions for my sister and me. What if
someone accidentally told us? Mom did not even tell her best friend, Aline, why
they had decided to leave Topeka, although Aline might have guessed.[3]

At the time of these events I was old enough to notice, but I have no memory of any dramatic change in the emotional atmosphere of our family or of any
questions that arose in my mind. Yet the emergence of such toxic information—
information that our parents believed should be kept from us kids—surely had
an impact on my sister and me. Keeping secrets requires emotional work on the
part of the secret holders, and children have sensitive antennae, picking up on
anxieties and fears without being aware of doing so. Consequential family secrets,
according to psychologists, can affect not only children's feelings and behavior
but also their cognitive functioning and performance in school. No doubt my
own later preoccupation with knowing and telling, perhaps even my early attraction to writing, derived in part from a childhood and adolescence of absorbing
the emotional fallout from my parents' worried knowledge of Huntington's and
from sensing their anxiety and guilt.[4]

My father decided that financial responsibility for the care of Mom's brothers
fell to him. None of them had secure incomes, insurance, or savings that would
carry them through years of disability. Jesse, the eldest, still worked in retail. Paul
and Seymour, the musicians, owned a bar and grill in Manhattan. But without

my father's help, they no doubt would have ended up in state psychiatric hospitals as their father had. Nor did they have children to assist them, perhaps because doctors had counseled them against procreation or on account of their own fears of passing on the disease, having witnessed their father's decline. Soon enough they would no longer be able to work. My father concluded that, to help them, he would have to leave Menninger, where salaries were low, for a more lucrative private therapy practice elsewhere.

Later in his life Dad came to believe that taking on these financial challenges helped him avoid worrying too much about the disease and what it could do to his own family. But I believe he had other motives, too: feelings of guilt about his relationship with another staff psychologist who had come to Menninger after the war to study diagnostic psychological testing with David Rapaport. Dad met Maryline Barnard before we moved to Topeka. Upon arriving for his job interview with Karl in the summer of 1946, Dad phoned his brother Henry, who drove over in a hospital car to take him to the Menninger Foundation office across town. "So I went outside and got in the car, and there was Hank sitting next to—Maryline!" Dad recalled. In his reminiscences Dad acknowledged frankly that he knew "right from scratch that Maryline was far more of a woman to my liking than Mother."[5] Maryline had grown up in Southern California on the land of the old Rancho Santa Margarita y Las Flores near Oceanside, where she had roamed the barrancas and wandered the hills with her sister and brother and helped her grandparents serve meals to the workers on her father's ranch. Tall, good-looking, and athletic, she had served in the Women's Army Corps during World War II and had female friends with names like Zelda and Peter, women who were as strong and independent as she seemed to be. Maryline was the West, the open bean fields and mesas blooming with lupines and shooting stars, a white Anglo-Saxon Protestant descended from stalwart early New England settlers, everything we were not. As Dad described her, "She was of the soil, granite when pushed, resilient and soft when that was needed."[6]

As he told it years later, he did not permit his relationship with her to interfere too much with what was going on at home. He preferred to "keep things stabilized just the way they were." At least he was honest about his duplicity. We came to know—and love—Maryline as a professional associate of Dad's who became a part of our family, a close friend included in many family outings and, later on, in family vacations and in the work on Huntington's disease. There were now two big secrets in the family, secrets my father must have worked hard to keep hidden and whose revelations would eventually transform us all.

FIGURE 5.1 At Menninger. From left: Martin Mayman, Robert Holt, Maryline Barnard, Milton, unidentified, and David Rapaport, circa 1948.

My father's decision to leave Topeka after five years was not unusual. Most people who came to Menninger, whether as graduate students in psychology, psychiatric residents, or staff members like Dad, stayed for a few years or decades and then moved on, some back to the cities they had come from, some elsewhere to new jobs. In a major exodus in 1948, David Rapaport, along with Robert Knight, Margaret Brenman-Gibson, Merton Gill, and Roy Schafer left Topeka for Stockbridge, Massachusetts, to take positions at a psychiatric hospital and research center called Austen Riggs. Henry and his family left the following year for New Haven, where his analyst, Albert Gross, had also gone. Bill Pious submitted his resignation letter to Will Menninger just a week before Dad did, telling Will he was leaving his "growing-up place." He was moving to New Haven as well.

Dad did not explain to Rapaport what "personal (family) involvements" made the East difficult to consider. New Haven or New York, near his and Mom's families and their closest friends, might have seemed the most logical destination, especially for Mom, who no doubt wanted to be near her brothers and Aline. (Sabina had died in 1947.) She found it hard to think of moving even farther away, though she tried to put a positive face on the situation. However, the fact that the East Coast analysts were firmly wedded to the idea of analysis as a medical specialty, requiring a medical degree to practice, meant that joining an institute and society and finding a hospital affiliation in New York or New Haven might have been impossible for my father, as it had been for Reik. Dad almost certainly considered that he was protecting Mom, too, and perhaps himself, from future obligations beyond financial aid. He may have wanted to get far away from both his and Mom's families—and from close involvement with Huntington's.

Mom's letters at the time alluded to the anxiety that she and Dad felt in the spring of 1951, politically as well as personally, with the Korean War raging and a growing national paranoia about Communist infiltrators and Soviet spies. It was as if the world would soon come to an end. And in a sense, her youthful world had come to an end, with all three of her brothers developing the disease that had taken the life of their father. For the previous twenty-one years, since Abraham's death, she had been able to put Huntington's chorea behind her. Now it was starting all over again, with not just one brother but all three. It is difficult for me to imagine how this avalanche crashing into her life must have felt to my young mother, just thirty-six years old, with two small children. And could she have noticed a change in her husband's behavior toward her, ascribing it perhaps to the pressures of his work so that she would not have to face yet another catastrophe? Just when she had finally begun to feel at home and make friends in Topeka, she was going to move once more, with the terrible knowledge of her brothers' destiny and one step closer to her own.

My father's decision to leave Menninger had one unexpected reward: a growing epistolary intimacy with David Rapaport, who took it upon himself to act as Dad's long-distance adviser in the matter of finding a place to settle and people to meet. In late March 1951, Dad and Sibylle Escalona traveled together to San Francisco and Los Angeles, where they both had been invited to lecture and meet with local analysts. Dad felt immensely grateful to Rapaport for his steady stream of effusive "Rapaportian" advice and support during this time. He was especially grateful for his introductions to Hanna Fenichel ("just about the most wonderful person there ever was"), the widow of the psychoanalytic theorist, author, and cofounder of the Los Angeles Psychoanalytic Society and Institute, Otto Fenichel; the lay analyst Frances Deri (whom Dad thought was "magnificent beyond words"), and the larger-than-life Ralph "Romi" Greenson ("whom I more than liked"), three prominent members of the Los Angeles psychoanalytic community. On his return to Topeka, and after much back-and-forth

FIGURE 5.2 David Rapaport, circa 1946.

correspondence with Rapaport, my father decided that Los Angeles offered the best possibilities for a private practice with hospital privileges and a welcoming psychoanalytic community. He felt that if he could establish good relationships with just Fenichel, Deri, and Greenson, the rest would not matter.

Though being at Menninger conferred an identity that remained throughout life, leaving Topeka was almost always traumatic, often bringing on a period of prolonged mourning.[7] Leaving for private practice—with or without a hospital affiliation—was a move Karl Menninger especially frowned on. Dr. Karl regarded private practice as commercial and somewhat greedy, out of sync with the social values and missionary spirit of Menninger. The Menninger community had long suffered from the tensions between commitment to a humanitarian and research enterprise on the one hand and commitment to an institution that needed to support a large staff of employees on the other. Salaries were low compared to the income available to analysts and psychiatrists elsewhere during the post–World War II years when psychoanalysts were sought after for the psychiatry departments of major medical schools. My father wholeheartedly embraced the Menninger ethos and its collaborative spirit and insisted he would not have left under normal circumstances. He always recalled his departure as deeply painful, akin to the end of a love affair, how Dr. Karl had refused to give him the farewell party traditionally held for those leaving, refused even to say goodbye, as if his departure were somehow a betrayal.

In this context, my father especially valued Rapaport's aid, concern, and even criticism, expressed in letters that were considerably more frank and direct than might have been expected given the reigning heterosexual culture of masculine stoicism and reserve. "Milt, I think that success and recognition have come to you deservedly if judged by your human qualities and ken," Rapaport wrote shortly before we left. "I think, however, that they have come to you relatively too easily, if judged by the amount of specific experience and theoretical foundation. Very soon financial success will be added to these successes and recognitions. These are seductions. I simply want to express my belief and hope that these seductions will glance off of you and leave you unmarred and the way I have known you." Rapaport knew he was being presumptuous and possibly even incorrect, "and yet I did not want to spare myself writing to you just the way I did."[8]

Far from thinking him presumptuous, Dad welcomed Rapaport's critique as an expression of intimacy. He was concerned only "that you [Rapaport] will not maintain your interest sufficiently to remind me every once in a while about the different meanings in success. I am much too dissatisfied with my work and my knowledge to feel very blown up about the kind of recognition I have received. But if you ever observe that I act as if I knew exactly what it was all about, then I wish you would jump on me with two feet and I will take it as a very great kindness."[9] He confided that he was really not able to tell Rapaport "how much I appreciate what you have done in making my path easier. It is a matter of very great personal importance to me that you have, on a number of occasions, expressed your interest in what I am doing."[10]

What we were doing in the summer of 1951 was going to Norway for two months. Trygve Braatøy, whom Dad had gotten to know at Menninger, had persuaded my father to come to Norway to treat a patient diagnosed with schizophrenia who was now locked up in an Oslo psychiatric hospital. Dad in turn convinced the patient's family, wealthy shipowners, to underwrite expenses for all of us, and off we went across the Atlantic aboard an elegant old Norwegian passenger ship, the *Stavangerfjord*, a thrilling adventure for Nancy and me. While Dad went to the hospital each day to see his patient, Nancy, Mom, and I roamed around Oslo, where we learned a smattering of Norwegian with kids in the local parks and where Nancy, with her golden hair and angelic appearance, often passed as a native. Dad took time off for us to go on a memorable road trip to see the spectacular Norwegian fjords and wander through strangely beautiful medieval wooden-stave churches with their special porches reserved for lepers, an architectural feature that impressed me greatly.

In early fall, at the height of the Red Scare in Hollywood, we landed in Los Angeles. Recent international events had strengthened U.S. fears of communism and "ushered in a period of political repression such as this country had not seen since the years following World War I."[11] This was a time when the FBI, or the specter of the FBI, surveilled everyone even remotely suspected of left-wing political views and when liberalism and even suspicions of same-sex attraction were equated with anti-Americanism and sometimes treason. A blacklist was growing throughout Hollywood that included the names of writers, actors, and even set designers and elevator operators. Teachers and many others also found themselves blacklisted. McCarthyism was in full swing. A second round of House Un-American Activities Committee hearings had taken place in the spring of 1951, just six months before we arrived, dividing the Hollywood community and destroying careers and, in some cases, lives. Such was the toxic political atmosphere we entered in the fall of 1951, ready to start anew.

Meanwhile the local analysts had recently split into two institutes on account of what they considered irreconcilable differences, both theoretical and practical. The original institute, the Los Angeles Psychoanalytic Institute (LAPSI), embraced the European refugee analysts who arrived in the 1930s and 1940s, some of whom did not have a medical degree but who stayed close to the theories of Freud and his daughter Anna. They were the "orthodox" Freudians who followed Freud in their acceptance of lay analysts. The newer group, the Institute for Psychoanalytic Medicine of Southern California, adhered more strictly to the medical requirement for psychoanalytic training but tended toward greater theoretical openness and eclecticism. Yet there was considerable overlap between the institutes, and some of the differences were more personal than professional.[12]

Despite these tensions, Mom presented a rosy picture to Aline, reporting many excursions to "mountains, oceans, parks, the desert," dinners, and parties and a generally active social life.[13] My father immediately had a full schedule and was turning away patients. The more unavailable he was, he joked, the more sought after he became. One contribution to his popularity may have been the enthusiastic reporting of a young actor named Robert Walker, the star of such popular films as *See Here, Private Hargrove* (1944), *Thirty Seconds Over Tokyo* (1944), and *One Touch of Venus* (1948), among others. Walker had been devastated by his breakup and divorce in 1945 from the actress Jennifer Jones and was eventually persuaded to seek psychiatric help at Menninger, where he became a patient of my father. When he returned to Los Angeles six months later, Walker felt he was a changed man. "I didn't want to go [to Menninger]," he told a reporter. "I thought it was a horrible shame to go to a mental clinic. I thought the way many people do—that it meant something like an insane asylum. Now I know differently."[14] According to his son, Walker had been reluctant to participate in any kind of therapy, "but Milton kept at it, and finally Dad started opening up to him."[15] Walker told the gossip columnist Hedda Hopper that he had carried around "completely needless burdens of self-doubt, hidden feelings of shame and guilt and so forth." He had been deeply unhappy. He thought everyone hated him. But the six months of "exhausting, albeit exciting, introspective research" at Menninger changed his view. "Most of what I learned of myself and [my] mental processes was in the nature of an emotional appreciation, not an intellectual knowledge," he told her.[16] He wanted to talk about his experience to reach people who might be suffering as he had and who could be helped by therapy. "If Bob has any messages," wrote Hedda Hopper, "it's to help others find peace of mind as he has."[17]

While we were still in Topeka, Walker returned to Los Angeles, continuing his therapy with Frederick Hacker, a controversial Hollywood analyst who had his own psychiatric clinic (and who had tried to persuade Dad to join him when

FIGURE 5.3 Newspapers all over the country carried stories about Robert Walker's enthusiasm for the therapy he had received at Menninger.

he heard my father was leaving Menninger). Walker made his best film during this time, the brilliant *Strangers on a Train* (1951), directed by Alfred Hitchcock, with a screenplay by Raymond Chandler, in which he played the part of an unforgettably sinister villain. However, on the night of August 28, 1951, Walker began drinking heavily and apparently became incoherent to the point that his alarmed housekeeper called Hacker, who immediately came to the house. Hacker had previously given Walker injections of sodium amytal to calm him down, with no ill effects. But on this night, Hacker apparently failed to take into account both the drug's interaction with alcohol and the possible toxic buildup of the drug in Walker's body from previous injections. Walker stopped breathing. He could not be revived.[18]

We were in Norway at the time of Walker's death, which must have been devastating for my father. He saved newspaper clippings about his former patient, not about his death but about his life and his enthusiasm for psychiatry. I think Dad wanted to remember Walker at a happy moment of his life. When he died, Robert Walker was just thirty-two years old.

My father was reserving judgment about the situation in LA, he reported to Rapaport in October 1951. He was trying to begin work slowly, with just four patients, to "see if I can really create a civilized way of life for myself, with some chance to read and write. I've been told that it's impossible in private practice," he added, "but I'd certainly like to try." Already, though, he had some misgivings. Sometimes he felt "a little depressed by what I have seen and felt in the social and professional atmosphere." As if trying to reassure himself, he added that he had been so busy that his judgment could easily be faulty and that he had not had time to see much of anything so far.[19]

As the analysts in Topeka had been, the analysts of LA—at least those my father befriended—were a social bunch who got together not only for dinners and parties (there was a "party of the month" club for a while) but also for amateur string quartets and play readings. Presiding over one of the LAPSI salons was the charismatic Romi Greenson (born Romeo Samuel Greenschpoon in Brooklyn in 1911) and his warm and vivacious Swiss-born wife, Hildi. Romi was a trim, compact man with a close-clipped, graying beard who kept in shape by playing tennis and golf until his first heart attack in 1955 at the age of forty-four, after which he slowed down somewhat. He had been a practicing psychoanalyst in Los Angeles since 1934 and a training analyst since 1947. Outgoing, emotional, and a compelling lecturer, teacher, and writer, Romi had the gifts of a stage actor and the comic skills of a stand-up comedian. To those he liked—and he soon came to love my father—he was a man of great generosity and charm. But he could be rude and outrageous toward those he did not like. During World War II, Romi had served in the U.S. Army Air Forces Medical Corps, stationed in Arizona and Colorado, where he had been friendly with Leo Rangell, now also prominent in LAPSI. But as Romi and my father became best friends, Rangell evidently felt excluded from his former intimacy with Greenson, leading to painful consequences for all three.

But all that came later. Throughout the 1950s, an eclectic group of friends gathered almost every weekend at the Greensons' welcoming Mediterranean-style home atop a hill on Franklin Street in Santa Monica. Many were analysts, but there were also actors, anthropologists, philosophers, historians, political scientists, musicians, writers, a dentist, and other intellectuals and their spouses. Visiting celebrities were always passing through, some related to the Greensons or their friends, such as the writer Leo Rosten, married to Romi's sister and the author of the comic novel, later made into a film, *Captain Newman, M.D.*, based on Romi's experiences in the army. At the Greensons' you might meet, as I did, the anthropologist Margaret Mead, now middle-aged and far from her adventurous youth in New Guinea, holding forth grimly from a wheelchair by the fireplace with her broken leg sticking straight out in a cast, or the actor Celeste

Holm, coolly beautiful and blonde, standing near the back of the couch with a cocktail in her hand.

I liked going to the Greensons'. They treated us kids as if we were grown-ups. They never talked down to us. But as the years went by, Dad began to chafe under the feeling of being more an audience member than an active participant at these gatherings. Later on, he became uncomfortable with the inclusion of some of Romi's celebrity patients at small Greenson family gatherings and outings (though not at parties or other large social events). The most famous of these was the actor and reigning Hollywood sex symbol, Marilyn Monroe, whom the Greensons adopted as a sort of orphan daughter, hoping to surround her with the warmth and comfort that had been so lacking in her early life. Dad was impressed by how intelligent and well informed she was, how articulate and direct in her speech. He did not get to know her well, but he would sometimes check up on her at her home when Romi was out of town, as Romi had requested, to make sure she was OK.

Romi was by most accounts a talented and empathetic therapist, despite his self-dramatizing and perhaps self-deception in relation to Marilyn Monroe. He was also an influential teacher and writer. *The Technique and Practice of Psychoanalysis*, published in 1967, was the first of several volumes that would become standard textbooks in psychiatric and psychoanalytic training courses.[20] In the late 1960s and early 1970s, Dad, Romi, and Maryline would share offices in the penthouse of a medical building on Roxbury Drive in Beverly Hills, with a view north of the Hollywood Hills. You could just make out the Hollywood sign on rare days when the smog lifted and the air was clear.

Dad had not been feeling well and was finding his work with patients with schizophrenia a strain, especially in private practice where he had little of the institutional support he had enjoyed at Menninger.[21] His family responsibilities had also increased considerably, as Nathan and Mollie had moved to Los Angeles around the same time we did and were living in the Mid-Wilshire neighborhood, close to the Jewish district surrounding Fairfax Avenue. Though a few distant cousins also lived in the Los Angeles area and visited the old folks from time to time, Dad mostly managed the household himself, with financial contributions from Henry. He often felt overwhelmed. Coming home from his office, he would eat dinner quickly with a frown and retreat to his study with its enormous wooden desk to dictate letters and work on the many papers, seminars, and presentations he gave throughout the decade, and for other reasons, too, as

FIGURE 5.4 Romi Greenson at his fiftieth birthday party, 1961.

we would learn later. Nancy and I were welcome to interrupt him in his study, but we were often so busy with schoolwork, friends, pets, after-school projects, practicing the piano (me), ballet lessons, and all the other activities of privileged teenagers in 1950s Los Angeles that most of the time we just took his retreat for granted. That's just the way it was.

Adding to my father's distress his relationship with Hanna Fenichel had soured, over what I do not know. However, the rift was serious enough that when Rapaport came to Los Angeles in the spring of 1954 to give lectures and get together with the Los Angeles analysts socially, it threatened to distance Dad from Rapaport as well. The ensuing tension led to a series of revealing letters between Rapaport and my father that illustrate the challenges of emotional expression between men in the 1950s, even between psychoanalysts. When Fenichel hosted a reception for Rapaport and invited my father, he attended but felt so uncomfortable that his behavior led Rapaport to think that perhaps Dad had soured on him. Dad in turn felt that Rapaport had acted coolly toward him. In subsequent letters, Dad tried to explain how all social contact with his colleagues at that moment was painful. "I suppose you must know that the friendly relationships that had existed with Hanna Fenichel are very much a thing of the past," Dad wrote to Rapaport afterward, apologizing for his silence following the Los Angeles meetings. He had gone to the reception hoping that Hanna had put aside "some of the feelings that have made social contacts so painful if not impossible. But apparently that was not so. And I suppose my discomfort made it seem that your own greeting to me was something less than warm." Dad apologized for his misjudgment. "I hope you will understand that towards you I have the strongest feelings of friendship and even gratitude."[22]

The ever-frank Rapaport accepted Dad's apology but also expressed some dissatisfaction. He noted that Dad had not contacted him during his visit, had not communicated with him, and had not commented publicly or privately on his presentation, "and that begins to be more than a question." Adding his analysis of the situation, he acknowledged that he had "no right to know and no inquisitiveness to want to know more about this. But I feel obliged to point out the disproportion between the cause you suggest (and I fully believe) and the effect." Rapaport hoped that they could continue their relationship but was clearly hurt and puzzled by Dad's behavior.[23]

In reply, my father revealed a vulnerable side of himself that he usually kept well hidden, as he put it. "You aren't fully aware of the inner emotional climate from which such effects spring," he told Rapaport a few weeks later. "In the last few months I have felt very hurt, very angry, and altogether too sensitive about many things in this atmosphere. Also, there are unsettling things of a personal nature." What things Dad did not say. He continued, "If I behave badly toward you (and others too) and find it difficult if not impossible to detail all the reasons, you ought not on the one hand assert no wish to intrude on my private affairs and, on the other, point up how inadequate is the explanation." Dad explained that he had imagined that Rapaport's greeting to him had been "very cool" and was not sure if it really had been or if he was imagining it. But he remained "tremendously uneasy and uncomfortable" the whole week of Rapaport's visit. He insisted that he would have liked to greet Rapaport warmly, to get together with him, and so forth, "but then he would be forced into a world of uncomfortable feelings, and details . . . which would bore you and leave any present moods and views more exposed than I care for them to be." He concluded the letter by saying that he hoped the matter was "well buried, or that future events will prove it should be."[24] Not yet ready to let the matter go, Rapaport pointed out "that there may be something not only in the 'climate' but in your handling of relationships (for instance, in this case the one to me) which you might want to look into—since to my mind no climate could, without something like that (whatever it is), produce such a discrepancy between cause and effect." After that, he was ready to leave the matter "well buried."[25] Dad evidently did not take offense at Rapaport's presuming to analyze his motives and psychology, no doubt feeling that such presumption was another expression of intimacy.

By the following year he and Rapaport—and possibly even Hanna—were back to their previous closeness, evidently without lasting damage. "Hanna just told me of your mother and I am terribly sorry to learn you have some new stress and sorrow to meet," my father wrote on hearing of Rapaport's mother's death. He continued, "I believe you are no stranger to pain, especially recently. I wish it were otherwise. I wish I could make it otherwise—but I guess grief and pain we generally suffer alone." Rapaport expressed appreciation for these empathetic words, telling my father that "they felt very good to me. It is still difficult to say more."[26]

My father continued to develop his thinking about schizophrenia in his ongoing dialogue with Rapaport. In one sense this relationship continued the male intellectual camaraderie he had established at Menninger with Bill Pious, a colleague with whom he could theorize and speculate. Now Rapaport became his interlocutor, someone far more knowledgeable than he was, against whom he could test his ideas. At times their dialogue grew so abstruse—brimming with incorporations, introjections, identifications, cathexes, and counter-cathexes—that even Rapaport, the great advocate of metapsychology, feared "this all may sound like hobble-de-gook."[27]

By 1953, my father had had turned away from his prior emphasis on a brutal archaic superego as the principal villain in schizophrenia, perhaps as a result of the critique and conversations at the Yale symposium and at other meetings where he had presented Nedda's case. Interpreting Freud's and Pious's formulations in a somewhat different way, he now emphasized an impoverished ego that had suffered a loss of emotionally charged (cathected) internal object representations—mental images or memories of people primarily. This is what Pious had described as "a sudden vacuum in the patient's psychic apparatus" and what Rapaport called the erasure of the road map. Going beyond their arguments, Dad now proposed that this emptiness or vacuum, this loss of a road map, this absence of mental images, somehow interfered with the ability to perceive reality, to distinguish between what was imagined and subjective and what was "real" and accessible to others.

With regard to this connection he began citing a line from Goethe in his talks and papers: "If the sun be not within us, how then shall we see the sun?" According to Dad, "It is as if he [Goethe] intuitively recognized that a process of internalization must precede the process of perceiving reality, or, as Freud put it, 'The process of testing reality is not to discover an object in real perception corresponding to what is imagined, but to rediscover such an object, to convince oneself that it is still there.' (Negation, 184)." Both Goethe and Freud seemed to affirm that "a sense of reality is dependent on the capacity to internalize objects and make stable identifications. The clinical counterpart of this idea is certainly to be found in the fairly characteristic feelings of desolation and emptiness complained of by many schizophrenic patients."[28]

He reflected on these issues in a long letter to Rapaport in March or April 1955. He had "spent the entire day (it is now late evening) puzzling over, chewing over, and digesting the seventh section of 'Organization and Pathology of Thought'" (this was Rapaport's paper "Toward a Theory of Thinking"), as well as the seventh chapter of *The Interpretation of Dreams*, the most metapsychological section of Freud's text.[29] Rehearsing Rapaport's explanation of how an individual's first love objects in the real world (usually one's mother and father) become templates for their first mental representations, images, or memory traces in the mind, my father then asked what would happen "if these internal objects . . . were suddenly to disappear?" He then acknowledged that the idea of "absence" was itself unclear: "Absent in what sense? Repressed? This is hardly absent." But then he imagined two circumstances in which the question might make sense. First, what if there were "some sudden repression of what may be easily accessible, quite conscious representations"? He wrote,

> Let me give a clinical illustration of a personal nature which may have some application. If I give a highly unsatisfactory discussion, express some silly idea, or the like, I am apt to be quite devastated. Afterwards I have the greatest difficulty in imagining I have a friend in the world, merit any respect, or have ever had a single intellectual success. Some singularly successful counter-cathexis is applied

to the memories of people or events where affection or other rewards were mine. A fairly common neurotic reaction, I believe, and one which I offer not to soften your criticism of my ideas but because I believe it has some pertinence.

He then described a second possibility, one that was "entirely speculative and may, indeed, be quite far-fetched. What if there were a sudden, massive withdrawal of cathexis from these inner representations and not merely the application of counter-cathexis?" he asked. "Would we not have inner world destruction comparable to that experienced by the too sudden, too traumatic withdrawal of cathexis of outer world representations? . . . Isn't the terrible aloneness, loss of identity etc. of the schizophrenic somehow related to the more or less sudden disappearance (absence) of these internal objects?" Although he insisted that "my thoughts are far more complex than what I have expressed here though not necessarily more sensible," he felt confident about his conclusion: "What I have said above leads me to [the] consideration of schizophrenia as a deficiency disease, highly variable in its manifestations, very different in prognosis when it occurs in childhood, adolescence, or maturity, but basically involved in the same typical memory defect, the absence of all-important inner objects."[30]

He did not claim that his arguments about the disappearance of internal objects or his emphasis on deficiency—which he articulated here for the first time—were new. He saw his propositions as extending Freud's 1911 analysis of the Schreber case and his 1915 essay, "Negation," in which Freud stressed the interconnection between the stability of internal objects and the ability to recognize reality in the external world. He also credited Pious's descriptions of an "abrupt emptying" and a "sudden vacuum" leading to the disappearance of ego functioning.[31] He was feeling increasingly assured of his views, but he still wanted Rapaport's approval. "You know, of course, that you are the only one with whom I can share such thoughts as these," he told Rapaport, adding that he had not found in Los Angeles anyone to take Bill Pious's place. "Besides I respect your judgment on these issues above everyone else so I find it more rewarding to try to express my floundering around to you." Despite "all the tortuous, half-informed, and uncrystallized thoughts" he had on these subjects, he felt that he had arrived at a conceptualization that accorded with his clinical experience. He would continue to talk of schizophrenia as a deficiency disease for decades to come.[32]

Rapaport was not convinced.[33] Trying again sometime later, my father offered a slightly different explanation. The "object representations are never lost, only not

cathected," he wrote in the summer of 1959. "More than that, I would call attention to the specific qualities of these object representations. They are fragmented,
partial, primitive residues of what existed earlier." As if anticipating Rapaport's
criticism, he proposed that "it is at least a good framework for observation and
even if I must make radical corrections in the theory, as surely I will, it is a start
for me toward expressing many other thoughts which deserve the light of day."[34]

My father never did persuade Rapaport. "To my mind the fragmentation,
the partial, the primitive residues are insufficiently explained by 'introjects' and
partial objects," Rapaport wrote, "because it is these very fragmented, partial,
and primitive objects which are integrated by the synthetic function of the ego
normally. To my mind it is the synthetic function of the ego which has suffered
primarily." In Rapaport's view, the problem was the ego's inability to integrate
all the various internalized images and objects—introjects, incorporations,
identifications, memory traces—into a stable and coherent identity. Nonetheless, the ever-enthusiastic Rapaport urged my father to write again so they could
continue the discussion. More important than mutual agreement was keeping
the dialogue going, with all its warmth and expressions of mutual appreciation.
"Schizophrenia is a mystery, and I am sure that your observations and thought
latch into the core of this mystery somehow," Rapaport wrote. He agreed "that
it makes no differences what kind of thought patterns organize the observations
for the moment. If it is a good enough thought pattern to stimulate observation
and to let facts slip by, then it is good enough."[35]

These are the last extant letters to and from Rapaport that I could find among
my father's papers or in the David Rapaport Papers at the Library of Congress.
Rapaport died thirteen months later, in December 1960, at the age of forty-nine.
No one ever took his place for my father, who loved him for his intellectual brilliance, his theoretical mastery, his confidence in Dad, his personal warmth, his
responsiveness, his emotional expressiveness, and what Dad called "his playful
mind." In Rapaport's death, Dad lost his most important mentor, the one colleague with whom he could be most fully himself.[36]

6

Losing the Road Map

It is easy to see the beginnings of things, and harder to see the ends.

—Joan Didion, *Slouching Toward Bethlehem*

In one photo, my ten-year-old sister sits behind me eating an orange in our outboard boat on Lake Tahoe while behind her Dad steers, smoking a cigarette and scowling into the camera, probably at Mom taking the picture. In another, Dad treads water, this time smiling up at the photographer standing on our dock—probably me—an unusual expression for my father at the time. I had just turned thirteen and gotten my period a few months earlier, an event I greeted with enthusiasm, according to the diary that I was keeping religiously. That summer I spent six weeks at a camp in the Trinity Mountains north of Tahoe and had a boyfriend for the first time. It was good to be away from the family, on my own.

The year 1955 was an especially difficult one for my father. In April, Romi suffered a heart attack, rendering him unable to see patients for several months. Dad took responsibility for Romi's most severely ill patients, returning to his colleague the income he would have received for their treatment, an act of generosity that neither Romi nor Hildi ever forgot.[1] So perhaps it was with a heightened sense of mortality that, the following summer, he decided to purchase, almost on a whim, a knotty-pine A-frame cabin in Carnelian Bay, on the north shore of Lake Tahoe, high in the Sierra Nevada mountains, complete with a boathouse and a dock we shared with the neighbors. For the next several years, we spent

FIGURE 6.1   Dad at Lake Tahoe, circa 1956.

a month or six weeks of the summer there and sometimes a week or two in the spring. Maryline, who had left Topeka for Los Angeles around the time we did, bought a house down the road where she and my father set up offices together. They had begun working together as co-therapists in Topeka and continued that pattern in California, including at Lake Tahoe where they could see the two or three patients who came during the summer to continue their treatment. These patients did not socialize with us and mostly kept to themselves, renting cabins near us on the lake accompanied by a family member or hired aide.

Maryline did socialize with us, spending many days and evenings at our cabin. A black-and-white photo from around 1957 shows my mother and Maryline

FIGURE 6.2   Leonore and Maryline on the porch of our cabin in Carnelian Bay, Lake Tahoe, circa 1957.

sitting on the porch: Mom is upright and tense, smoking a cigarette, while Maryline lounges seductively in the chair next to her—a portrait of our family life. We swam, took picnics, made excursions, and waterskied behind the boat that Dad bought that first summer on the lake. Rapaport, who suffered from heart problems, was never able to visit. Nor was Theodor Reik, from whom Dad had gradually grown more distant. But Bill and Jess Pious came, the Greensons too, and many others. Visiting analysts hatched plans for a workshop on schizophrenia on our dock. And there were Nancy's friends and mine. It could have been paradise.

My father continued to wrestle with his ideas about the nature of schizophrenia and to suffer from what he called an inner dissatisfaction with the unclarity of his thinking in so many areas. "I am afraid I do not really work well alone," he had confided to Rapaport a few years after we came to LA. "It is a flaw in my make-up that I must be pressed by outside forces in order to be productive. Unfortunately the kind of people with whom I communicate readily and who would push me to more creative work are not around here." He was very fond of a few people and respected them greatly, "but I am not quite able to use them as the cuttlebone to sharpen my own wits." He blamed his inflexibility and "a certain reticence with those who are closely anchored to established knowledge and theory and foreclose me from daydreams and fantasies."[2] Perhaps, too, his status as a lay analyst excluded him from university contacts that might have been intellectually rewarding. UCLA stood "very strongly against psychologists with therapeutic pretensions," he noted in one letter, alluding to a faculty member who had said that he would feel "contaminated" by any professional association with Dad.[3] My father would soon begin to collaborate with Romi Greenson. But Romi, for all his brilliance and imagination, was neither the clinical conquistador that Bill Pious had been nor the Talmudic scholar that was David Rapaport. Dad missed them both.

Could he have achieved more as a theorist and thinker if he had been able to interact with them more closely and more regularly and been part of a less contentious psychoanalytic community away from the temptations of Hollywood? At Menninger, he had been regarded as a brilliant guy, a major intellectual figure who was going to make important contributions, which analysts who were psychiatrists (and thus MDs) typically did not expect from PhDs. He seemed suited for a position of leadership.[4] Even Karl Menninger wrote twenty-five years later that he regarded Dad "as one of the greatest psychotherapists in the world and I don't see my young colleagues following in your footsteps at all." But perhaps he was always going to be more of an activist, "by nature involved," as he told Nancy years later. These are questions I am certain he asked himself, as I ask as well, without finding answers.[5]

In the late 1950s, a thirty-six-year-old professor of psychiatry at UCLA, Philip R. A. May, initiated a major research project at Camarillo State Hospital in Southern California to compare different treatment modalities for patients with schizophrenia in a hospital setting. Chlorpromazine (Thorazine), one of the first antipsychotic drugs, had come onto the U.S. market in 1954, and the study aimed to compare the effectiveness of drug therapy to other forms of treatment.[6]

My father, who headed the Research Division of LAPSI at the time and who maintained his enthusiasm for research, was one of nine psychoanalyst consultants on the project (and once again the only PhD among them) along with eight (non-analyst) clinical psychologists, four psychiatrists, seven nurses, eight social workers, and thirty-seven residents in psychiatry who performed most of the psychotherapy.[7] May's team selected 228 individuals hospitalized for schizophrenia for the first time at Camarillo between 1959 and 1962, male and female in roughly equal numbers, and apportioned them among five groups. One group received individual psychotherapy alone for a minimum of one-hour and an average of two-hour sessions two or three times per week; their therapists were mainly the residents in psychiatry under the supervision of senior psychoanalysts with experience treating patients with schizophrenia. May defined the therapy given as "in general ego-supportive and reality defining," with a minimum of depth interpretation, a focus on helping patients address their immediate life situations, and an emphasis on the therapist acting as a "suitable model for introjection." A second group of patients received antipsychotic drugs alone. A third group received both individual psychotherapy and drugs. A fourth group was given electroshock but no other form of treatment. A fifth group, considered a control group, received what they called "milieu" therapy—another legacy from Menninger—which was defined as basic ward care and staffing at a level considered "good to superior for a public psychiatric hospital" at the time. It included nursing care, occupational and hydrotherapy, recreational and industrial therapy, but not psychotherapy, either individual or in groups. Treatment lasted one year.

May and his colleagues published the results in a series of papers from 1964 through 1966 and in a book, *Treatment of Schizophrenia: A Comparative Study of Five Treatment Methods*, which appeared in 1968. As principal investigator, May summarized the outcomes: drugs alone and drugs plus psychotherapy were the most effective treatments and the least costly in terms of time and money. However, investigators found little clinical difference in the outcomes of those who had received drugs alone and those who had received drugs plus psychotherapy, which was significantly more expensive to administer than drugs alone. Psychotherapy alone and milieu treatment alone were the least effective and most expensive, while electroshock occupied "an intermediate position." May also concluded that "by almost all the clinical, movement and cost criteria, drug had a powerful beneficial effect, while the effect of psychotherapy was nonsignificant."[8] Indeed, May was emphatic that for most patients hospitalized with schizophrenia, the addition of psychotherapy to drug treatment provided relatively small gains, if any. He proposed that what he called psychotherapeutic management might be more worthwhile in helping patients adjust to life outside the hospital.[9]

Dad was deeply unhappy with the report. In his view, the results confirmed the outcome of only the limited form of psychotherapy employed in the study, psychotherapy given for a mere hour or two twice or three times a week by relatively inexperienced practitioners. He worried that the results as the published papers and book presented them would lead to unwarranted generalizations about the futility of psychotherapy in general for treating schizophrenia rather than demonstrating the apparent ineffectiveness of the specific limited psychotherapy used in this one investigation.[10]

Although psychiatry was moving increasingly toward a reliance on drugs to treat mental ills of all kinds, my father continued to receive invitations to lecture and write about the practice of psychotherapy for schizophrenia as well as about his theory of schizophrenia as a deficiency disease.[11] He consistently tried to link his abstract theoretical ideas to the felt experience of his patients. "Locusts may ravage the lands, floods destroy the seed, fire destroy the crops; but we are poor historians to say that locusts, flood, and fire killed the people. Starvation is the ultimate evil in such events," he wrote in one paper.[12] He emphasized the desolation, the sense of dying and of being dead, the devastating aloneness, the feeling of inner catastrophe experienced by some of his patients. He described their sometimes bizarre appearance: "walking like a grand lady, like a slinking bawd, like a little child, or like a tiny shadow who would rather not bother anyone at all."[13] And he continued to call for an intensive therapeutic approach aimed at creating a stable therapeutic relationship—object constancy—to make it possible for the patient to "internalize" the therapist as a step toward developing a more secure and stable sense of self.

Dad composed most of his psychoanalytic papers as presentations for professional meetings, symposia, and conferences. They were essentially published talks, with the characteristic rhythms and repetitions of his speaking style. "This is the true meaning of the emptiness and fear to be seen in the schizophrenic. This is the ultimate meaning of the stilted, stiff, and awkward movements. This is what destroys the sense of a unified identity," he would say.[14] He also elaborated his ideas in the seemingly endless lectures and seminars he gave at universities, medical schools, hospitals, and veterans' centers around the country and at LAPSI. His lectures often dealt in detail with specific therapeutic techniques and situations, such as the first meeting with the patient, approaches to the family, and dealing with separations caused by the therapist's travels. He spoke of not overpromising, not being too friendy or eager, meeting patients where they were and not where the therapist would like them to be, and putting into words feelings that patients communicated but could not express verbally as a way for the therapist to show that they were understood: commonsense practices in which my father excelled.

He also began collaborating more formally with Romi, although they had been exchanging ideas informally since they first met. After fifteen or sixteen years of working together, my father noted quizzically that they had "mutually infected each other, and I think it is rather difficult to say where my ideas end and where his begin."[15] This cocreation was especially evident in an influential paper, "The Non-transference Relationship in the Psychoanalytic Situation," which they presented at the 1969 International Psychoanalytical Association Congress in Rome. They challenged the idea that all responses of the analysand to the analyst were necessarily transference reactions, that is, unrealistic and inappropriate emotional responses based in the past rather than reality-based reactions to the analyst in the present. They argued that the analyst and the patient were "two real people, of equal adult status, in a real personal relationship with each other" and that analysts should attend to this "real relationship" and be respectful of the patient's life experience and cultural background, which may differ significantly from their own. "There is always the possibility that we, rather than the patient, are unrealistic about the external world, including the patient's world," they wrote. "Technical errors may cause pain and confusion but they are usually reparable; failure of humanness is much harder to remedy."[16] Such thoughts might seem unremarkable to those outside psychoanalysis, but Dad expected that the Rome paper would kick up a furor. "The stick-in-the muds of psychoanalysis will be highly indignant; the radicals will be equally skeptical," he wrote me. "So we will probably stand quite alone, perhaps joined by those mediocre talents who have no place to go anyway and will be glad to join up with any homeless leaders."[17]

Au contraire, quite a few in the audience agreed with my father and Romi, with much of the discussion centering on such semantic questions as what constitutes an interpretation and the meaning of the "real." There seemed to be little disagreement with my father's call for building on what was positive and healthy in the patient, as well as for humanness and humility on the part of the analyst. Dad's expectation that he and Romi would stand in lonely isolation went unfulfilled.

By the time of the Rome congress, psychoanalysis had begun to lose its cultural cachet in the United States. Activists in the social movements of the sixties attacked psychoanalysis as conservative and bourgeois, especially feminists, most of whom saw it as a bulwark of patriarchy.[18] My father evidently remained outside the generational debates and discussions of "common ground" that roiled the congress, taking little notice of the shifting centers of psychoanalytic interest from the United States and Western Europe to Eastern Europe, Asia, and Latin America, where younger intellectuals were drawing on psychoanalytic thought to deconstruct legacies of colonialism, racism, and authoritarian rule. While psychoanalysts still dominated medical school departments of psychiatry in the United States, younger psychiatrists entering the field were more biologically oriented, focused less on the mind than on the brain.[19]

The reasons for this biological revolution in psychiatry are complex and debated by historians of science. Ann Harrington has argued that the shift represented less a response to new discoveries about the brain and behavior, which were still limited, than a new vision of psychiatry's identity and destiny. Joel Braslow has suggested that the emptying of state hospitals and the lack of funding for alternative treatment in the community helped push psychiatry toward an increasingly narrow reliance on drugs.[20] At the same time the emergence of neurochemistry and discoveries related to brain chemicals (neurotransmitters) such as serotonin and dopamine, along with new imaging techniques, helped inspire brain scientists in the sixties to create the broad interdisciplinary field of neuroscience, a word that first appeared in the *New York Times* in 1965, signaling the start of a transformation in the way Americans understood themselves. Research was revealing that experience was embodied in the brain and that memory produced physical transformations throughout the body; mental acts were not exclusively of the mind. The new Society for Neuroscience, founded the same year as the Rome congress, held its first meeting in 1971, with 1,395 scientists in attendance; within a few years annual meetings would morph into gatherings of many thousands. New drugs such as Prozac coming onto the market in the eighties hastened the move of psychiatric thinking away from psychoanalysis and into the fields of neurobiology and genetics.[21] Although many psychoanalysts at the time were indifferent to, if not disdainful of, biology, as Eric Kandel has described, it may be that those few analysts who worked with patients with schizophrenia, like my father, had always been more open to biological explanations than was the profession generally. For even in the 1950s, quite a few shared the view that the etiology of schizophrenia would ultimately be understood in terms of physical and physiological factors.[22]

In this changing milieu, Dad continued to participate intermittently with colleagues in discussions of research on the psychotherapy of schizophrenia, perceptions of which were also changing as the disease became what Jonathan

Metzl has called "a protest psychosis," a diagnosis increasingly given to Black men and associated with violence and danger.[23] In 1971, my father joined a small working group, one of five around the country, organized by the National Institute of Mental Health (NIMH) in response to the limited success of available drug treatments in improving the long-term psychological and social functioning of people living with this disease.[24] The NIMH project aimed to bring experienced clinicians together to develop proposals for clinical research. Their mandate was to isolate variables that could be rigorously tested, such as the benefits and limits of specific therapeutic techniques (similar to what was assessed in the Camarillo study), the involvement of social workers and psychiatric aides in treatment, methods of matching patients to appropriate therapists, the characteristics of patients most likely to benefit from individual psychotherapy, and the qualities most essential in a therapist for such patients.[25] The issue of whether the loss of stable internal object representations—about which everyone seemed to agree— should be considered a deficiency or a defense (against anxiety) was also on the table, and my father thought it fundamental to treatment considerations. He thought these working groups would be promising for correcting "all the misguided, inept, and faulty questions which have dominated the field of thought about schizophrenia, including some of the disastrous misconceptions which seem to flow out of the Camarillo study," as he wrote to Hussain Tuma, one of the working group participants. "It will take a long time to recover from the current distortions which may emanate from Phil May's book."[26]

From his travels and contacts abroad, Dad now proposed that therapists look overseas for alternatives to U.S. treatment modalities for people with schizophrenia, to Japan, Norway, and the Netherlands, where social workers and doctors visited the homes of patients who had recently become acutely psychotic. He thought clinicians should collaborate with cultural anthropologists on cross-cultural studies of schizophrenia treatment to come up with hypotheses for further research. But with a few exceptions, he was no longer seeing such patients and no longer deeply invested in schizophrenia research. He had come to see individual psychotherapists working with severely disturbed individual patients as "serving no great social purpose except as their knowledge is extended broadly to general populations. Otherwise, we lead a rather parasitic existence."[27]

In 1981, he had one more chance to immerse himself in discussions of schizophrenia when he joined Nancy, an administrator at what was then called the National Institute of Neurological and Communicative Disorders and Stroke in Bethesda, Maryland, for a traveling workshop in China sponsored by the Jennifer Jones Simon Foundation as part of its focus on mental health.[28] The group of nine included Herbert Pardes, an eminent psychiatrist and psychoanalyst who had left a position as chair of the psychiatry department at the University of Colorado Medical Center in Denver to become the director of NIMH and would soon become Nancy's life partner. Dad was probably excited as much by the chance to travel with Nancy as he was by the opportunity to visit psychiatric hospitals in China. And despite their hosts' lack of interest in psychotherapy, he found much to admire. He applauded their practices of trying to keep patients at home, emphasizing education and structure, mobilizing the extended family and the community to help, and stressing the importance of "a patient's place in society, his obligation to society, his duty toward society." He also praised their insistence on cleanliness and fostering social interaction to help preserve a sense of dignity and build self-esteem. Dad believed that this firm approach strengthened patients' sense of identity and increased their perceptions of reality. He may have startled his Chinese listeners when he claimed that Lenin's argument that "the human brain knows the world because it mirrors objective reality" was roughly equivalent to Freud's view of the ego as "a precipitate of abandoned object cathexes." But I imagine that they appreciated his praise of their treatment approaches, which he called superior in certain respects to those in the West.[29]

What remains of my father's years of dedication to studying, treating, and lecturing on schizophrenia and his efforts to theorize the metapsychology and therapy of the disease? Does anyone read his papers anymore? What of his seminars, discussions, and reviews of his colleagues' work, his talks to general audiences? Is his work at all relevant today? I asked Robert Wallerstein when I interviewed him at his Belvedere, California, home in 2014. "It's historical," he told me. "Most people coming into psychiatry and psychoanalysis in the twenty-first century will never have heard of Milton Wexler."[30]

As his daughter, and someone who is neither a psychoanalyst nor a historian of psychoanalysis, I am not in a position to evaluate my father's theory or practice. I can say that he was one of a relatively small number of psychoanalysts who took on the exceptionally difficult challenge of treating patients with schizophrenia with psychotherapy—not psychoanalysis per se but definitely talk therapy—in the days before medication became widely available and who documented his struggles. His clinical case reports, provocative theoretical proposals, seminars, wide-ranging discussions with colleagues, and enthusiasm for research garnered the respect of many eminent people in the field. Those whom he mentored as far back as Menninger recalled him fifty years later with admiration as someone they wanted to emulate. And more so than many analysts, he welcomed a dialogue between psychoanalysis and the emerging field of neuroscience, with its emphasis on empirical evidence, even if by the late 1960s he was moving in a different direction. As psychiatry turned increasingly to drugs, he and a few other analysts kept alive the value of talk therapy when combined *with* drugs and social support, even if they disagreed on precisely what that therapy and support should be. Most of all he upheld the therapeutic benefit of a strong relationship between therapist and patient, whether one accepted his idea of "internalizing the therapist" or not.[31]

As the law professor and mental health advocate Elyn Saks wrote in her 2007 memoir, *The Center Cannot Hold: My Journey Through Madness*, "Medication has no doubt played a central role in helping me manage my psychosis, but what has allowed me to see the meaning in my struggles . . . is talk therapy. . . . There may be a substitute for the human connection—for two people sitting together in a room, one of them with the freedom to speak her mind, knowing the other is paying careful and thoughtful attention—but I don't know what that substitute might be."[32]

7

Freudian Fathers and (Proto-) Feminist Daughters

Psychology may have constructed the female, but it also helped to construct the feminist.

—Ellen Herman, *The Romance of American Psychology*

When I was a teenager, my friends sometimes asked me how I managed to be so normal since the kids of shrinks were known to be crazy. In truth, I did not feel normal. But it was only much later that I began to wonder how Freudian ideas, especially ideas about sex and gender, may have shaped my sister and me since Freud was almost a god in our household. Dad did not interpret our dreams, watch us anxiously for neurotic behavior, or interpret our imperfection as a narcissistic wound. I never felt under surveillance, as some children of psychoanalysts did.[1] If we spilled a glass of water at the dinner table, he did not look at us worriedly to sense what unconscious motivation this so-called accident might reveal. He was not strict, but he set definite limits, and crossing them brought some kind of reprimand or mild punishment. I remember an occasional spanking, but these were so rare that they stand out in memory. Mostly he was protective rather than punitive. "I don't care what everyone else is doing," he would say firmly when we wanted to join friends on an outing of which he did not approve. "You're *my* child, and you're not going!"

Dad's philosophy of child-rearing and his approach to treating schizophrenia had much in common, as he spelled out in his first paper on Nedda, when child-rearing must have been much on his mind:

> Some time ago my mother objected to the fact that I wished to make a certain recommendation to a physician who was treating a schizophrenic patient in whom my mother had a strong personal interest. The recommendation was simply that the physician should actively put some brakes on the negative, hostile, tempestuous acting out which characterized the patient's behavior at the time. My mother's feeling was that such behavior could be dealt with only by sincerity, kindness, love, understanding, a host of other verbal shibboleths which, though valuable as abstractions, say not a thing about the kinds of behavior which will communicate such sentiments.
>
> "Mother," I said, "how would you feel if you were suddenly made governor of New York State?" "Why, I'd die of fright," she said. "And that," I said, "is what it is like for an unprepared and weak child to be made governor of the household with unprecedented power to gratify its wishes. It isn't love but fear, and sometimes hatred, which makes us abandon the child (and the patient) to his own emotional devices. You must sometimes hold the baby, and hold it tight, or it will be too much frightened by the rage and lust inside."

Criticizing the permissive strategies of Paul Federn toward his patients with schizophrenia, Dad argued that controlling approaches could be reassuring. Imagine, he suggested, if an angry, rebellious child threatens to run away from home, and his father "quietly sanctions his leaving. Not for me and not for my children! I would not grant them such terrifying power over their environment at the moment when their most frightening and uncontrollable impulses threaten expression."[2]

More than forty years later, he couldn't help chortling over a *New York Times* report that he felt confirmed his beliefs. "I am not in favor of a spanking culture," he wrote to "sundry parents" in a brief memo in 1996, "but neither am I in favor of a lollipop and ice cream culture." Dad claimed that in the sixties, when children learned how to blackmail their parents with drugs or by dropping out of school, he had seen many instances of "family values crumbling into the dust." The idea that families should be a democracy seemed to him a total disaster, and he was pleased when research seemed to confirm his belief. He welcomed a renewed emphasis on old-fashioned values "which for all their stuffiness, [seem] now to have been much more effective than the subordination of parents to the domination of their children."[3] Still, he was not as conservative and certainly not as

authoritarian as he sometimes sounded. In one lecture on the therapy of schizophrenia, he alluded to the qualities of a good parent that he thought useful to
the therapist: guidance, suggestion, support, love, anger, interest, and hope, "all
the parameters very far from good analytic technique." In another, he defined the
good parent (and therapist) as "aware, unafraid, unashamed, unhurried, tender
without seduction, firm without tyranny, critical without destructiveness, per-
ceptive without cynicism or laughter." With Nancy and me, he had all of these
qualities much of the time. One quality, though, eluded him. "I do not think a
good parent hides his feelings," he wrote in a 1958 lecture. But for a long time,
hiding his feelings was his method, a way of being he could not escape.[4]

Throughout the fifties Dad tried to give Nancy and me the firm guidance he
and Henry felt they had lacked growing up, as well as the freedom to explore
our interests. Fatherhood in the post–World War II United States, particularly
among the white middle class, had become an increasingly valued role, a source
of identity and pride for men. As the historian Rachel Devlin has shown, much
popular culture of the fifties encouraged an intimate, even eroticized relationship
between middle-class fathers and their daughters. Far from being more independent, teenage girls in the fifties were in some ways closer to their fathers than ever
before. That was true of Nancy and me.[5]

At the same time, our father was so busy that for days we did not see much of
him. He often came home after we had eaten dinner with Mom. Or if he joined
us, he was tired and preoccupied. I learned only much later of the extraordinary
scope of his activities in addition to seeing patients all day. Or perhaps I was not
paying attention. He was collaborating with Bruno Bettelheim in Chicago on a
project relating to children with autism. He was giving lectures on schizophrenia
at the Los Angeles Psychoanalytic Institute, the University of Minnesota, the
Jewish Children's Bureau in Chicago, Napa State Hospital, the USC medical
school, and the Veterans Administration hospital in Los Angeles, among many
other places. He was serving on committees at both LAPSI and the American
Psychoanalytic Association. He was active within the California branch of the
American Psychological Association, working to secure licensure and insurance
reimbursement for clinical psychologists against the wishes of most analysts and
psychiatrists, who thought only *they* were qualified to practice therapy.[6] He was
director of the Training School and head of the Research Division at LAPSI,
and co-taught, with Romi Greenson and Leo Rangell, a LAPSI extension class
on the psychotherapy of schizophrenia. He was giving a tribute to Theodor Reik

FIGURE 7.1 The fifties. Nancy and me, circa 1959.

on Reik's seventieth birthday, defending the value of struggle against conformity, apathy, and injustice. He was trying to patch things together at home and keep his emotions under control.[7]

Throughout the fifties, I identified completely with Dad's thinking about every-thing except, maybe, Elvis Presley and rock and roll. I did not question his impa-tience with Mom or why Maryline came to so many family dinners and outings at Lake Tahoe during the summer. Ours was a patriarchal family par excellence, with Dad not so much a benevolent dictator as a charismatic chief who ruled through the eager submission of his women—all of us held willingly in his thrall (including our Weimaraner, Sheba). In a letter to me many years later, Maryline recalled dinners at our house during which Mom, Nancy, and I all hung breath-lessly on Dad's every word, while she was the only one who dared to speak up and offer an independent, or even contradictory, idea. "You would be overjoyed that I would dare to take the opposite point of view from your father," she wrote to me in 1986, "would dare to say something that did not agree to the letter with what he said. You felt that as very daring and would listen gleefully. At that time the three of you (Mother, Nancy and you) would only listen silently, adoringly, worshipfully at whatever he said or ordered. True he was seldom wrong, but I was not afraid to stick my neck out and you enjoyed it when I did."[8]

Patriarchy seemed to me not only natural but desirable. I understood James Dean's anger and despair in *Rebel Without a Cause* (1955) when he sees his father down on his knees scrubbing the kitchen floor and pleads with him to stand up, take off his apron, and act like a man. *My* father wouldn't be caught dead washing the kitchen floor! Nor, I thought, would most of the analysts or other men with whom our parents socialized. More than anything, I wanted to be the kind of woman my father liked, but I could not quite figure out what that was; although a schizophrenogenic or refrigerator mother was one kind of monster one must avoid becoming at all costs. The messages at home were confusedly mixed. On the one hand, Dad encouraged a certain resilience and tough-mindedness in Nancy and me. He urged us not to depend on the approval of others, to think for ourselves, to get an education, to consider a serious career. On the other hand, he was surrounded by women who deferred to men, starting with Mom, who resembled the ideal "feminine" woman lauded by the psychoanalyst Helene Deutsch in *The Psychology of Women* and by Marynia Farnham and Ferdinand Lundberg in their scurrilous *Modern Woman: The Lost Sex*, a best-selling screed against all things feminist sitting on Dad's bookshelf. Yet he appeared to prefer Maryline, who did not seem lost and did not match this ideal at all. But even Maryline, with her competence and self-confidence, positioned herself as Dad's subordinate, his associate as he called her, even if she was more assertive in her day-to-day stance. Except for Hanna Fenichel and one or two other female analysts, those few professional women in our parents' circle who seemed truly independent, like the biographer Fawn Brodie, he considered "aggressive" and therefore flawed as women.

Of course most of Nancy's and my friends looked to peers, not parents, for approval and ideas of how to act. They looked to the movies and television for models. Parents represented everything they did not want to be. But we daughters of parents with cultural cachet found ourselves torn. Who were we, mere kids, to stand up to profound psychoanalytic truths, especially when the ideas of Freud and his followers pervaded so much of popular culture? We loved Freud, too. As Devlin has written, "Embraced simultaneously by the psychiatric profession and by a postwar public eager for explanations of 'war neuroses,' psychoanalysis had a kind of authority during these years [during and just after World War II] that was unparalleled, not just in the history of psychiatry, but in the history of social-scientific ideas."[9]

As I learned later, psychoanalysis was open to many interpretations and Freud was far less doctrinaire about women and the feminine than many of his followers, including Reik, who claimed that "the biological and emotional divergences between the sexes are basically unchangeable and no mere alteration of our cultural pattern will ever reverse them." While Reik basically gave a psychoanalytic veneer to Victorian gender ideology, Freud took a far more progressive view. "It is essential to understand clearly that the concepts of 'masculine' and 'feminine,' whose meaning seems so unambiguous to ordinary people, are among the most confused that occur in science," he famously wrote in 1915.[10] And as Joanne Meyerowitz has shown, the popular culture of the fifties presented many images of women beyond the exclusively domestic, for example, of women participating in politics and in labor, peace, and civil rights activism. Even Gidget, that fifties icon of blonde, blue-eyed, teenage femininity, wanted to surf with the boys. Alternatives were out there. I just did not know where to look.[11]

More and more Dad's face wore a scowl around the house. Mom's growing sadness and withdrawal cast a pall over our home life. Caught up in my own adolescent turmoil, I had little interest in what was happening to my uncles far away on the East Coast. When the youngest, Seymour, came to Los Angeles for occasional visits and made us laugh with his jokes and magic tricks, winking and twitching all the while, I chalked it up to Mom's peculiar family, whom I usually wrote off as "weird." No one mentioned Huntington's chorea. Like many parents in their situation, Mom and Dad worried that talking about it with Nancy and me would only cause us worry and fear. They rarely spoke of it with their friends, or with each other for that matter. In fact, they barely spoke to each other at all. At times Mom seemed more like a servant than a spouse, doing all the drudgery and boring chores. At other times, she resembled another daughter. She didn't seem to have a voice of her own.

Mom was clearly suffering from depression, no doubt mourning the illness of her brothers, whose future she knew only too well and whom she saw perhaps once a year. She carried within her the weight of her terrible secret and having no one to talk to about it, or so I believe. What did it cost her, I wonder now, to hide that knowledge from Nancy and me all those years? And what if she suspected that Maryline was something more than her husband's colleague but felt too frightened to find out? Possibly her depression and passivity signaled the subtle onset of Huntington's or her recognition that it was beginning. Years later Dad would chide himself for not having recognized early symptoms. Yet, in retrospect, who can say if Huntington's symptoms or psychological responses to her bleak situation lay at the heart of her withdrawal?

From the outside, we appeared a successful fifties family. But inside, it was all so confusing. I concluded that if the silent atmosphere of our family was normal in a marriage, I would avoid a similar fate by not getting married and not having children. I felt inchoately that if Dad did not love Mom, then she was not lovable, and therefore neither was I—since, as everyone told me, I looked just like her. I came to feel that becoming a wife was a terrible fate, equivalent to a kind of living death. I was determined to avoid it at all costs, even if I did want love, romance, and sex with a man. My rejection of conventional feminine roles also extended to motherhood, which I imagined as another kind of enslavement. I did not want to relive through my children the unhappiness of my own teenage years and could not imagine a different experience. Years later Nancy and I wondered aloud to each other whether our parents, especially our father, had conveyed to us—consciously or unconsciously—the idea that we should not have children because of the possibility of Huntington's. That eugenic perspective still prevailed among knowledgeable physicians as late as the 1970s. As one authority wrote, "There is probably no other disorder with such a strong argument against reproduction." In his view, the fact that people at risk did have children suggested that "human reason and human emotions move on separate tracks, rarely accessible to each other."[12]

By the late 1950s, many things were changing in our family life. Increasingly Dad spoke of his parents as if they were his patients or children. Mollie's anxiety and increasing paranoia and Nathan's irritability and hot temper made their everyday life a constant challenge, which also weighed on Dad. He was always driving across town to "settle them down," as he put it. I remember him talking about tussles over food, for my grandmother, growing ever more suspicious, suspected the household help Dad had hired of poisoning them and sometimes refused to eat, whereas Grandpa Nathan, a large man, liked rich Jewish food all too much. Dad reported to Henry that the bland food prescribed by Nathan's physician was contributing to their father's depression. "His first response to my suggestion that I would get somebody to cook for him was 'It will have to be a nice Jewish woman,'" Dad wrote to Henry. "His idea of diet and Uhley's [the doctor's] will probably be very different."[13]

I graduated from University High School in June 1959 and entered Stanford University in Palo Alto in the fall, leaving my sister to negotiate the painful parental situation on her own. Two years later I went even farther afield, leaving the country altogether to spend six months in Tours, a city south of Paris, as part of a Stanford-in-France program. Little did I know that the family I had left behind in Los Angeles would not be the family to which I would return.

8

Revelations

Listen, you guys, my great excess was I wanted to live.

—Saul Bellow, *Henderson the Rain King*

I first heard about the possibility of a divorce in the summer of 1962, when Mom came to Paris, where I was staying after the sojourn in Tours came to an end. While ensconced with thirty-nine fellow students in a former Tours hotel in the middle of the Loire Valley, I had not paid much attention to events unfolding at home. As I learned later, Dad had begun considering a divorce late in 1961, as his twenty-fifth wedding anniversary approached, and he realized that he could not continue in a marriage that had long been barren.

Years afterward he admitted how unhappy he had been and how "if I hadn't gone to my study every night I'd have run away or killed myself. I certainly would not have had the strength or the mood to listen to you or Nan. It was my safeguard." For endless years he had lived with that empty relationship, he told me, "and that is the only thing you could sanely accuse me of doing that may have been damaging."[1] For a long time I was angry with him, not so much for seeking a divorce as for not getting divorced sooner, for the "atmosphere of dislike" that he could not hide, and for the sense of betrayal I felt when I finally learned the truth about Maryline. My anger touched a raw nerve in him, in part I believe because he recognized its legitimacy. On the other hand, I did not then appreciate how much Dad was struggling to find a way out of his unhappiness without abandoning Nancy and me or hurting us all too much. (Years later he told me that one of his themes with patients was the importance of honesty in relationships because he was so conscious of the ways that he had been dishonest, trying to hold things together.)[2] I was still trying to find my bearings two years later and must have written to Romi, because I found a letter he wrote to me about the divorce, offering his advice and what he considered his support. In his view, the divorce should not have been so startling since it had been obvious for many years that Mom and Dad were not happily married. The reason it was so painful for me, he declared, was that I had been living in a sheltered world, a dream world, and had not been willing to face reality, including the reality of my father, whom he considered "a remarkable man and also remarkably complicated." He continued, "I think his idea of what warmth is [is] very different from most people's, including mine. He is a worshipper of thinking, knowledge, ideas, and talk, but he has little regard or capacity for expressing tenderness, affection, and sentiment. The only exception to this is for people who are completely helpless and desperately in need; then he is wonderful, but the more mature you are the more likely your father will be intellectual, perhaps witty, even friendly, but without any physical demonstration of tenderness or affection."

Romi assured me that while he was a strange man, my father was "no monster. He is also not the greatest man in the world." Romi admired him very much, "but he is very difficult. However for me he is very rewarding. I do not think any child of his would have an easy time of it unless the child were severely crippled." I do not recall what I made of this startling letter at the time, but rediscovering it more than fifty years later, I find it more revealing of Romi's ambivalence toward Dad and his desire for more affection from him than insightful about either Dad or me, though he was probably correct in his assessment of my immaturity and that I needed to just grow up.[3]

In his book *Out of Sight: The Los Angeles Art Scene of the Sixties*, the art critic William Hackman describes 1962 as "the annus mirabilis for the Los Angeles art scene." It was the year that the history-making curator Walter Hopps moved to the Pasadena Art Museum from the landmark Ferus Gallery he had helped establish on La Cienega Boulevard in Los Angeles. Several major new galleries also opened on La Cienega that year, turning it into a major site for showing and seeing art. Although the New York art world had for a long time dismissed Los Angeles as a provincial outpost, by mid-century the city had a considerable heft of artists and a sprinkling of galleries. If it did not yet have an infrastructure of museums, collectors, and a press dedicated to art, it had excellent art schools.

These schools provided the foundation that would soon give birth to a flourishing art scene, one that my father would embrace.[4]

By the late fifties, young artists and curators were moving to LA in droves, many of them to attend the Otis College of Art and Design or the Chouinard Art Institute, forerunner of the adventurous California Institute of the Arts, better known as Cal Arts. A strong sense of place informed the new aesthetics, as Los Angeles became a mecca for the bold, the wildly inventive, and the curious. The sculptor Edward Kienholz and his friend Hopps had opened Ferus in the spring of 1957, showing the work of such young LA artists as John Altoon, Billy Al Bengston, and Ed Moses. Four years later, the Watts Towers Arts Center opened to support and show the work of Black American artists, with the sculptor Noah Purifoy as its first director. The following year, Hopps began showing contemporary art at the Pasadena Art Museum, including the work of Los Angeles "Finish Fetish," "light and space," and outsider artists such as the sculptor Miriam Hoffman, whose work my father collected. The Gemini G.E.L. artists' workshop opened in 1966, offering assistance to young artists who would soon gain international renown. In East LA, the Chicano artist collective called Asco, or Nausea, would later form in response to the mounting deaths of young Mexican Americans in the Vietnam War. The Woman's Building opened downtown in the early seventies to support and show feminist art. Although Dad may have already met a few artists, it was John Altoon, the brilliant, charismatic, but troubled artist at the center of the Ferus group, who initially brought Dad into the wider art world. My father's experience with schizophrenia was the thread that wove his initial connection to a community of young, mostly white, male, and macho artists—especially those associated with Ferus— whom he would cherish, nurture, admire, love, and learn from for the rest of his life.[5]

Stories differ on how they met. According to Ed Moses, Altoon in full delusional mode had climbed over a fence at the veterans' cemetery in West Los Angeles and was running naked among the graves until the police came and hauled him off to the state psychiatric hospital in Camarillo, some one hundred miles away. So Ed and his friend, the collector Laura Lee Stearns, called Dad who then drove with Stearns up to Camarillo to see Altoon and get him out.[6] My father, on the other hand, recalled that Altoon had gone berserk on La Cienega, going from gallery to gallery destroying paintings and sculptures. A few artists with work in these galleries finally managed to grab hold of him. Someone, possibly Moses, had heard that Dad was a therapist who specialized in treating schizophrenia.

FIGURE 8.1 Clockwise from left: Babs Altoon, John Altoon, Milton, and the art collector
Marcia Weisman, 1967.

Photograph by Ken Price.

They called Dad, who urged them to bring the offender to his office, where he
would try to calm Altoon down.

In the catalog for a 1997 retrospective of Altoon's work at the Museum of
Contemporary Art in San Diego, my father described his initial encounter with
Altoon, whom he felt had "slipped over the edge of the precipice but had not
fallen all the way." He perceived that Altoon still had some doubt about what
he claimed was his God-given mission to destroy all the art on Earth and then
teach little children to create art that was genuinely fine and noble. He was willing to negotiate. After they had talked for hours, Dad told Altoon that he must
return to his studio but promise to come to see him every day so they could
"reach some understanding of the heavenly will." To everyone's surprise, Altoon
kept his promise. Dad was eventually able to vanquish the delusions and keep
them away, at least most of the time. Many of Altoon's friends and his future
wife, Roberta Lunine (known as Babs), felt that Dad helped him produce his

best work by easing his demons, setting him free, and making him "part of the world, part of the rest of us."[7]

Through Altoon, Dad began meeting other artists and going to Monday night gallery openings on La Cienega, spending time afterward with some of them at their West Hollywood hangout, Barney's Beanery. He bought art or exchanged therapy hours for art. Altoon's drawing of witches pinching the nipples of naked, spread-eagled starlets appeared on the wall over Dad's bed. Michael Olodort's enormous painting of a sinister, grinning, upside-down Humpty Dumpty now loomed over the dining room table. The *Anxious Soldier*, Stephan von Huene's leather and wood sculpture of a man's leg topped by two testicles ensconced in an open briefcase, stood watch over Edward Higgins's three-foot-high bronze *Owl* sculpture on the living room floor. As more artists began asking for therapy, Dad started a free therapy group, although most members had individual hours as well. He drew up a proposal for researching creativity in hopes of getting funding to treat more artists for free.[8]

Years later Ed Moses recalled, "One day Milton called me up and said, 'How'd you like to get your head shrunk for nothing?' That sounded interesting." Ed joined the group and stayed three or four years. "Instead of whining and complaining, you'd talk about your fantasies," he told me. "Milton was always trying to get everyone to talk about their fantasies." The emphasis in the group was on "being free, being able to free associate." Ed recalled that since he was able to talk easily about his fantasies, Dad sometimes used him as a shill to get the group working. "But he could also nail me. He was fair but unfair."[9] He liked Dad's approach as a therapist because he talked, he told you what he thought. "He wasn't one of those shrinks who sat there silently," Ed told me. He had "perceptiveness and rationality." And with his law background, he was pragmatic; he offered practical advice, for instance about dealing with clients and for both professional and personal relationships. My father loved having artists for patients "because they brought to me original and exciting perspectives on the world. They not only saw the physical world in a unique way, but they also saw human relationships and behavior in unusual ways. In that sense they were bringing to me as much as I hoped that I brought to them. That was worth everything to me."[10]

Ed Moses introduced Dad to the young architect Frank Gehry, who would have a profound influence on Dad's life. Born and raised in Toronto, Frank had studied architecture at the University of Southern California, where he began to make connections with the painters, sculptors, and ceramicists on campus, as well as with those from Chouinard and Otis. Unlike most architects at the

FIGURE 8.2 Milton, Maryline, and Ed Moses: "He [Milton] was an explorer. He admired people who were willing to step out and try different things. And he was a person who would step out and try new things too." Circa 1976.

time, he identified more with these visual artists than with his fellow architects. Ed Moses and Frank Gehry became friends, and Ed suggested that Frank see my father to help him resolve problems in his marriage. Frank was reluctant, fearing that therapy might interfere with his creative process. But he went, and rather than advising him about his marriage, Dad told him to go home and come to a decision first; only then would he try to help. Which is what Frank did. Eventually he got married again, happily, to Berta Aguilera, a warm, vibrant, and accomplished woman from Panama with whom he formed both a professional and personal partnership. Besides individual therapy, Frank also joined Dad's artists' group, an experience he recalled as transformative. "Milton taught me to see yourself as others see you," he told me. He had spent two years sitting silently in the group, getting angry at times but not saying anything, "until one day they all turned on me and said, 'Why do you just sit there silent, being critical of everyone else?' I was shocked. Because I always felt that I was shy, that's why I didn't talk. I didn't realize that I appeared so angry and judgmental to everyone else. They thought I was sitting there criticizing them." After that Dad helped him deal with his anger, which Frank felt greatly improved his relationships with clients.[11]

Like Ed Moses, Frank had been worried about Dad—and therapy—interfering with his work. Dad's conventional aesthetic tastes—the dull midcentury modern

FIGURE 8.3   Frank Gehry and Michael Olodort's Humpty Dumpty, 1960s.

furnishings in his office, the sentimental paintings on the walls—raised alarm in his mind. "I'm all fucked up," he told Dad, "but it's not about that." He felt confident about his work and did not want anyone interfering. He recalled later that "Milton never crossed the line. He was not judgmental about my work." The connection between Frank and Dad morphed from a relationship between therapist and patient into one almost like father and son, deepened by my father's enormous admiration for Frank's genius as an artist and architect and his love for Frank as a wise and generous human being.[12]

While 1962 was a historic year for the arts in Los Angeles it marked painful events in my father's life, events both distant and close. In August Marilyn Monroe died. One night several months earlier, Romi had had a meeting to attend—or possibly he was in Europe—and asked Dad to look in on her. After work Dad drove to the house on Fifth Helena in Brentwood and found Monroe lying on the floor in a hallway, so drugged he could not wake her. "I could not even lift her," he told me twenty-six years later, when he asked me to write down his memories of Monroe. She had taken too many pills to sleep, pills from the prescription bottles arrayed at the side of her bed. He and her friend and acting coach Paula Strasberg, who was in the house but had been occupied in the kitchen and unaware of Monroe's condition, gave her coffee and walked her around to revive her. Dad also took away the pills at her bedside. Strasberg told him that Monroe had been anxious and upset that evening. She was filming a water scene the next day in which she was required to be nearly naked, and she did not want to do it. "I don't think this was a suicide at all," my father told me. "It's somebody who accidentally takes too much because she wants to sleep." He believed her death was "a tragic accident with a lot of tragic detail—people who used her as merchandise."[13] Dad was saddened by Monroe's death, but Romi was distraught, both at the loss of his patient and at finding himself the target of conspiracy theories surrounding her death. Romi's grief spilled over to Dad, adding to his sorrow.

Sometime that same tumultuous year, the year that the civil rights movement gained momentum in the South and the Cuban Missile Crisis brought the United States and the Soviet Union to the brink of nuclear war, Dad's close friend from Topeka David Rubinfine, now an up-and-coming analyst in New York, left his wife of many years, Rosa, and announced his intention to marry his famous patient, the writer, actor, and director Elaine May. The scandal cost him his position as a training analyst at the New York Psychoanalytic Institute and would scar his reputation for life. David's divorce affected Dad deeply, as he was contemplating his own, though their situations differed. He was concerned about David's future and his two young daughters, especially after Rosa committed suicide the following year. He and Henry also worried that David seemed to be exchanging one dependent relationship for another, something Dad wanted to avoid for himself. But he agreed with Henry that their friend had "the right to break with what is unhappy for him." It was as if Dad saw in David a kind of mirror and was trying to convince himself as well.

Most significant emotionally for both brothers in 1962 was the death of their father, Nathan. He had been diagnosed with prostate cancer many years earlier, but it had remained in remission until early in 1962, when it returned and metastasized to his bones, leaving him in excruciating pain. Whatever his reservations about his father's strength of character, Dad loved Nathan for his warmth and generosity. Perhaps, too, my father's consciousness that he had repeated the pattern of my grandfather's infidelity made him more forgiving than he had been in his youth. He was haunted for the rest of his life by a memory from Nathan's last weeks in the hospital when he needed more and more medication to keep his pain barely under control. At a certain point, when the meds were failing, Dr. Uhley asked Dad about increasing the morphine. Dad understood that this would soon cause his father's breathing to stop, but he desperately wanted to put an end to his father's agony. "You could almost hear the bones cracking," he would tell us with a grimace. Dad told Uhley to increase the morphine. Within hours, our grandfather died. Dad never regretted that decision, but it weighed on him, knowing that he had made the call that led to his father's death.

Freud once wrote that the death of a father is the most important psychological event of a man's life and his most poignant loss.[14] For it was after the death of his own father, in 1896, that Freud embarked on the self-analysis that led to his most significant work, *The Interpretation of Dreams*, using his own dreams as case material. And it was after my grandfather's death on March 6, 1962, that my father embarked on his own self-analysis ("not psychoanalytic," he insisted) and took the first steps to change his life. That summer he attended a psychoanalytic conference in Mexico City—a city he had never been to before—and took Nancy with him. Afterward he went fishing and scuba diving off the coast of Mexico. He also began writing a series of soul-searching letters to Henry, unlike any he had written before. He and Henry had always been close, but now, in the aftermath of their father's death and my father's impending divorce, their communication became more intimate. The Mexico sojourn was a revelation, he told Henry after his return, a watershed moment, "one huge, exuberant exercise in creative living for me and maybe for Nancy. By contrast, many of our family activities, our group living, became painfully dull and stereotyped; mainly because [of] an atmosphere of tension and dislike that had to be dealt with." The last summer at Tahoe had been "sufficiently painful to make all of us want to give it up." But scuba diving in Mexico had been transformative. "It was a clean wash in a new world," he wrote. "It was just a symbolic look-see down in a new environment, freed from previous conceptions since I hadn't experienced it before. It was even wonderful to be disoriented about distances and find that I couldn't reach the things with my hands that seemed so easily reachable." He had wanted to go scuba diving, "not because I like that sort of thing or want to pretend to be a kid. I wanted merely to get off this solid earth and be unencumbered by any rules or regulations, even those of gravity, space, time. I wanted to feel as comfortable standing on my head as on my feet. That's a lousy illustration. But it does have something to do with what I want for my life. I haven't got it. I probably never will have it. But I'm trying."[15]

He continued articulating thoughts to Henry that he rarely shared, admitting that he did not know in what direction he would go, but he knew he could no longer "stay put." Dad thought his brother had misunderstood a remark he had made in a previous letter:

Not the big prick; not the big name; nor the fame and fortune . . . When I speak of something big these days I mean only one thing: the Big Freedom. I mean the escape from bondage. I mean the loosening of all ties with what is phony respectability and consecrated but moldy values. I don't any longer think very ambitiously about anything except grabbing up what remains of life and commanding it to give. And what I want it to give is the ability to be as deeply honest, as deeply independent, as deeply indifferent to the fraudulent modes that bind us as one can attain. I can't be too specific. It's not something that lends itself too easily to words. It's a kind of feeling that has been developing in me more and more and has become urgent.[16]

Henry was skeptical. "The Big Freedom is an interesting idea and a more interesting yearning," he wrote to Dad, "and it is something I have wrestled with for a number of years now." He characterized Dad as "a man of the times, so to speak. It is the existentialism that our social philosophers of today have been preaching in various ways and which has given rise to loose sorts of movements, as the Beatniks, the Kerouacs, or the more isolated 'respectable' types. If it were not so prevalent today, I might think you are beginning to suffer the climacteric [sic] desperation. If so it is well rationalized." Henry wondered "if the B.F. isn't one

B.I. (Big Illusion). You yearn like the bound Prometheus for freedom (as so many have); one reason he never achieved it is that at the last he didn't know what that really was. But I do not decry his effort to obtain it nonetheless—and if scuba diving or space flights will seem a road towards it why [not] take it and more power to the journey."[17]

The following week Dad sent off another long introspective letter to his brother. He agreed that giving in to the irrational was not the answer. Responding to Henry's effort to distinguish between rationality and conventionality (while admitting that conventionality can sometimes be rational), my father argued that neither he nor Henry had "managed the distinction much as we hope[d] to. I tell you honestly that I lead a conventional life (more or less) and so do you."

> And I try all the time to pretend that it is a rational life and based on rational aims. That, so far as I can presently see, is all bullshit. You wouldn't stay in your present position for five minutes if you were really rational. Nor would I. And that's the rub. We do lead dull, stereotyped, and conventional lives and try to pretend that our work, the exploration of psyches, the dedication to family, to ideals, makes it all rational. Bullshit again. We are just too damned afraid to tear loose from the traps we built for ourselves. We are guilty and timid. And that is the damned truth I learn more and more of every day.[18]

For Dad, something more than a rejection of conventional values was needed, a "middle ground" that involved "the attempt to see things for what they are, divorced from prior stereotypes, infantile attachments, etc. Perhaps this is only a simple-minded way of saying what analysis ideally aims at—and almost never achieves. It means successive approximations to a truth which is probably never really achievable but which I, at least, remain too far removed from. It means making new aesthetic, moral, and practical judgments without reference to the 'sets' established by all the sticky and mainly irrational elements in past history." He felt that "something more is needed than merely hating sham, pretense, or cruelty. We all hate or say we hate these things. Something more is needed than new moral or ethical codes, however rational these may be." But what? "We lead bullshit lives," he continued. "And what do you mean you can't get out of it? Do you mean anything else than that you haven't the courage to do what you really want to do?" He questioned his own courage as well. "I'm fed up to the ears with my present arrangements in life," he told Henry. "Lee bores the hell out of me. And if it weren't for the shadow of dread which hangs over my life and her desperate inadequacy I'd get out.... [But] I may only just be a god-damned shit without guts also to do what I want and even should do and then [I] rationalize it all in the name of morality, conventionality, the kids, my patients, and all the other crap that keeps us prisoners." Looking around at his "conventional, upper middle class crap house," with its "mixture of *House and Garden* and *Home Beautiful*," Dad was filled with disgust. His home had "as much personality, as much individuality, as a Sears Roebuck refrigerator. And your house too, my friend." At this moment in his life Dad could hardly restrain his loathing. "And it's dead because our lives are dead. And our perceptions are deadened. And we are frightened and intimidated. And shits." And so were some of "our dear, dear analysts."

Still, Dad admitted that he did not know what he was going to do. It would "depend on what order of shit I really am. . . . I don't usually share these things as you know. But that is mainly because I doubt that others will understand. . . . And if you say there is no solution for any of us I will also agree. But then let us live in Kierkegaard's agonies, not in Babbitt's nirvana."[19] He was not attracted to the idea of giving "free expression to instinctual urges and contradicting all conventions for the sake of some illusory freedom. So toss that one. Nor am I interested in the directions pointed out—or lack of directions—by Kerouac, the Beats, etc. These are all negatives. They are nihilistic, destructive of feelings, immersion in death. They are against something and instead of rising in protest choose the path of indifference, withdrawal, and a 'cool' emptiness which seems too closely allied to schizophrenia to find me much attracted." Dad found Camus's existentialism compelling, but "I'd rather, at this moment, write and think in a different direction." He was looking for something to be for.[20]

One thing he was for was a continuing engagement with the psychoanalysts in Mexico City. At the summer 1962 meeting he had become friendly with an analyst named Alfredo Namnum and his wife, Marie. Afterward he and Namnum corresponded about integrating the Mexican Psychoanalytic Association with the West Coast psychoanalytic group—or even merging the Mexican group with LAPSI, although this plan did not come to fruition. But Dad made several trips to Mexico in the 1960s for conferences in addition to his trip with Nancy and stayed in touch with Namnum for several years. "I'd love to go to meetings down there," he told Henry. He was bored with the West Coast meetings and thought the Mexico group would liven things up.[21]

By December 1962, Dad was gaining new energy and optimism, even using an awkward new slang more reminiscent of a teenager than a middle-aged psychoanalyst. On the eve of a benefit premiere of a new film about Freud directed by John Huston, Dad told Henry that "it should be a real gone film since

Sartre and many others have had their finger in this pie and John Huston did the directing."[22] Dad had never talked like that before. He didn't talk like that later either, at least not to me. Reading his letters in 2019, I suspected this language reflected his associations with the young artists he was meeting. He was starting to have fun.

He began talking to Henry about the divorce settlement and about his enthusiasm for turning over most of his assets to Mom. He rejected Henry's suggestion that he was giving up too much: "I just don't want possessions, money, and caretaking roles for things. Lee finds it important to her safety. Great. Let her have it. I know it's not simple but I have some delight in the sheer recklessness of it; and it has very good reason behind it, I think, though you do not. I begin to feel freer already. So it's my form of pleasure and I let you label it what you like."[23]

Dad also took issue with Henry's notion that "the Hollywood climate, the babes, the plucking-fucking 'one or another for one's delectation'" played any role in his decision to leave his marriage. He scoffed at Henry's idea that he had "smelled so much bullshit out there that it is invading everything—that is, the smell of it."[24] As Dad put it to his brother, "There are women everywhere and those around here are no prettier than elsewhere, no looser, and no more interesting. I'll bet you've got your daisy chains and whore houses all over staid New England." What he really objected to were those who "screw the ass off anything in skirts and ditto with their wives to keep the old girl happy. Or make loving excuses when they are too tired. But they don't leave home; they need that as a shield for excursions and retreat." In one of his most impassioned statements, Dad summed up what he was against and what he hoped to find:

> I'm after fresh ideas, not fresh meat, though no objection to the latter. If I wind up with the sort of thing your letter suggests then I will have failed. And it is very possible, perhaps even probable, that I will in the larger sense. But leaving all of that aside, I couldn't continue here and as is. It's too damned dull and commonplace and uninteresting. Which leads to the question about creative living and creative activity. I suppose the distinction is not too marked but it does make some sense for me to separate these. I'd like to write more and I put that in the realm of creative activity. Creative living is more difficult to define since it is a much broader realm.[25]

Dad thought he knew ways to live creatively, and "working my tail off in the office, coming home to a somber atmosphere, meeting routine obligations, and accepting stupid invitations to stupider parties" was not what he had in mind.

He wanted to try something else: "And if I fall flat on my face; or some pretty witch wiggles her ass in four-part time and makes me feel longer and bigger and trips me up; or I get so damned lonely that all my dreams seem worthless; or I wind up sick and penniless and dependent on those [to whom] I've ceded my all; or I haven't got a damned thing to contribute worth using for toilet paper: then I've failed. But I've tried. I won't then have a sick feeling that I'm just a horse's ass using a bag of tricks to explain what a great guy I really am if only circumstances weren't against me."[26]

He still could not entirely explain what he was doing, even to his brother:

> Alice is not the only one who is dissatisfied with the answers I give. Everyone else is too. You too. I suppose it always seems unreasonable for someone to do what they want to do. It must be that they are pushed by some deeper motive. One doesn't make decisions in life because one is bored or disinterested. One must have a secret lover, special perversion, etc. Maybe my secret perversion is that I like to be free, to do what I please, and to pay any financial price for the liberty of escaping from total dullness. Maybe that is cruel, narcissistic, selfish, and the like. But I know what I can survive and what I cannot survive. And I do enjoy living. So I do what I must.[27]

He had no desire to give up his practice as a psychotherapist and psychoanalyst, the bedrock of his identity. Nor did he contemplate fewer responsibilities, either to Mom or to Nancy and me. Mainly he was reflecting with all of us about what he wanted in the years ahead; he was thinking with us about how to live.

Even as he was discovering the "Big Freedom," Dad remained very much in our lives, especially after Mom suffered a devastating assault and rape in the summer of 1963, in Mexico City, where she had come to rendezvous with Nancy and me after our sojourn studying Spanish in Guadalajara. We were not quite finished with our classes at the time, and Dad flew to Mexico City to bring her home and help her recover, asking Nancy and me to return as soon as classes ended. I am grieved to think how impatient I was with my mother when I came back to LA and how little empathy I showed in the days and weeks that followed, failing to grasp fully the depth of her trauma and the horrific reality of what had happened to her, a divorce followed by a violent assault that could have ended her life. Looking back, I think she was starting to show symptoms of Huntington's, for even before the assault she had seemed increasingly nervous and jumpy, traits I attributed at the time to her general anxiety and what I considered her inadequacy. And yet, after Dad moved out of our Napoli Drive house in March 1963 and even after the assault that summer, she took extension courses in education and biology at UCLA and in 1967 received a California credential to teach biology in junior college. She volunteered at a genetics lab at UCLA, attended Democratic Club lunches, went to art shows, and took painting classes. I was starting to feel proud of her.

By the time Henry and his wife, Liza, visited LA again, driving across the country in the spring of 1964, Mom and Dad were ensconced in their separate lives. "We visited Lee Wexler's new apartment," Henry wrote in his journal, "very lush. Milt had his own apartment—less lavish but nice." Henry reported Dad's explanation for the breakup in his journal: "'Hank, it's simple. People think there must be all sorts of complex reasons for such a thing. Nothing of the kind in this case. I just got bored. Fed up. Life was too dull. Now I feel alive, free, unfettered and uninhibited. Most people expect dramatic reasons; there are none.'" To a colleague from Topeka, the psychologist Robert Holt, Dad put it differently: "Being structured the way I am I enjoy single life and have no regrets on that score at all."[28]

Although I felt myself in the midst of a "Big Confusion," at some point I must have written Dad a Father's Day letter supporting his aspirations and dreams, for he wrote in response one of the most loving and open letters of his life:

Dearest Pooch,

I don't think you can know what a wonderful Father's Day letter you have sent to me. It gets to the heart of crucial questions and I am delighted to address myself to them. Not that I have answers but I have many thoughts and let me try them on for size.

In the first place I am really touched that you want me to kind of take off and live what I say I believe. I am inclined to think that is impossible but I don't intend to evade the challenge. But I want to challenge you first on one fundamental premise that I think may clarify the distance between our perspectives. You are constantly saying that I lead a life of "sacrifice, uninvolved, intellectual." There are times when I wish that were so but it is so far from the truth that I wish I could really erase that notion from your head. But it is very hard because seen from the outside, at some distance, it may look that way. To the contrary I lead a highly charged, very emotional, often sentimental, reasonably romantic type of life in which intellectual matters often go by the boards. I live far more by emotion than by intellect and I am not just simply referring to the dramatic day-to-day life with patients. In fact the greatest struggles of my life are involved in the effort to keep those disciplined intellectual forces functioning so that I don't get bowled over by feelings that often run away with me.

He suspected that I wanted him "to document that with a secret love life, hidden relationships, a personal diary, or a book of poetry." But I was going to have to take his word for it:

What you and a lot of other people see, even those quite close to me, is a very considerable reticence and a readiness to use my intellect to maintain some distance. It doesn't seem to occur to any of you that this may be contributed to by two things: a certain skepticism about a way of life and a way of thinking and feeling that doesn't either meet my needs or seem very meaningful to me; plus the possibility for which you (and they) never allow that there is a very real need in me for privacy since my thoughts, feelings, and even actions may be outside the mainstream of conventional rules and this is not always comprehensible to others.

Dad explained that he didn't particularly believe in confessions: "I'd rather respect the other person's rights to lead his life as he will and communicate what he wishes. All I can say is that it makes me laugh, sometimes a little bitterly, at the amazing preconceptions some people have that I am a stolid, cold intellect. There are a few people in this world who would also laugh heartily at any such notion. You also will change your mind in time. But you must be patient with me and perhaps a deeper truth will seep in with the passing of time."

Reminding us of his financial obligations to Mom, which would require him to continue working at an intensive pace, he allowed that he did want to take longer holidays and work a bit less. He needed the time to "develop more and more what I want to be radical about. You see that my present radicalism is altogether too negative. . . . I am merely balancing on tippy toes, trying to keep out of the shit, but not exactly knowing to the deepest and fullest what I really believe in." When I confided that my doubts elicited great anxiety, Dad said he felt it was "very exciting and even fulfilling to be in this doubt-ridden state." He knew that he did not want to be burdened with possessions, but that notion did not go far: "In the main I want to find a few values that I can really live for; or some person or persons; or some work; or some combination of all. I don't treasure either intellect or work as the whole answer. They have a place. But I will not accept a world for myself that doesn't have love in it; or passion; or brave, new thoughts that verge on brave new feelings." There had been a "gradual shift away from philosophy in the direction of aesthetics." He wasn't being a martyr in giving Mom nearly all his assets: "I am trying to buy my freedom, perhaps at any price." Reassuring me that he was willing to respect our differences and that he wanted to hear my thoughts and likes and dislikes, he concluded on a note of humility. "If there is one damned thing I have learned better than most (and I mean most) it is that the truth is harder to come by than to talk about. Happy dreams and many thanks for the Father's Day letter," he wrote. "It gave me courage all day. I think we'll both make it sooner or later."[29]

9

The Big Freedom

Adjustment, that superannuated and fatal conception for which an antidote must be found.

—Robert Lindner, *Prescription for Rebellion*

"Ulysses is home," Dad announced to Nancy and me in late October 1964 with a mixture of joy and sadness. "How did I ever wait so long to do what is so meaningful to me?" On his own for the first time in twenty-eight years, he had flown to Paris, his jumping-off point for a month-long road trip through France and Italy soaking up architecture and art. Nancy had started her sophomore year at Radcliffe, and I had just set off for Caracas under the aegis of a Fulbright fellowship to Venezuela. Mom, too, had begun her single life in an apartment Dad had found for her while he had moved into one of his own. In our different ways, we were all leaping into the unknown.[1]

In letters to Nancy and me over the next several years, Dad recounted his adventures while trying to help us navigate the changed realities of our family and of our individual lives. He was inventing his new way of living in these letters, which were a kind of journal and space for reflection. He did some of his best writing in this correspondence. I did not always appreciate the advice he offered, yet I saved all his letters, as if to keep him close to my heart, even during the angry times when, in my head, I was pushing him away.

"I found it just wonderful to live at the little pensions, take sponge baths, or be dirty, wash my own shirts, and in general keep my focus on the art," he wrote to us after that first European trip. He fell in love with Romanesque churches and despite his atheism admitted it was "possible to get religion in such an atmosphere." He felt peaceful, enjoyed the silence, and did not feel lonely, he told us, thanks to interludes of socializing along the way. He also met colleagues when he gave a lecture on schizophrenia at a medical school in Florence, where most of the psychiatrists who attended were opposed to psychoanalysis. Even so, he reported, he managed to win friends and influence people. They asked him back to give a seminar, and his host took him on a tour of the city, capped by a magical moonlit visit to cloisters designed by Michelangelo.[2]

I cherished the image of my father roaming through Florentine churches and through the Galleria Borghese and the Capitoline Museums in Rome, the city he loved most, pencil in hand, jotting notes in his blue-and-white Michelin guidebooks, which Nancy and I used later on our own forays in Europe. He had looked at a world "which in startling ways turned out to be too much like my fondest imagining. Reality should never approximate a dream, and when it turns out to have some identity with the fantasy it gives you a queer feeling," a feeling he would try to reproduce in a screenplay many years later.[3]

By the time Dad returned from his European trip, Nancy was ensconced at Radcliffe, and I was attending classes at the Central University of Venezuela in Caracas, a modernist architectural masterpiece designed by the great Venezuelan architect Carlos Raúl Villanueva. It felt good being in a city and country where neither of my parents had ever been, as if coming to Venezuela marked my independence. I was supposed to be studying "social change." Yet no one checked whether students attended class, demonstrations and strikes closed the university for days at a time, and we were pretty much free to do as we pleased. Feeling unfocused and bent on practicing my Spanish, I spent less time studying than socializing, sometimes with an older graduate student who became my boyfriend and guide on field trips with his anthropology professor through the plains of the southern Venezuelan state of Apure. In Indigenous villages we observed haunting all-night healing ceremonies, smoked a mescaline-like drug called *yupo*, and interviewed Yaruro Indians, who were more interested in securing clean water, schools for their kids, and modern health care than in discussing the details of their traditional practices.

I suspect that if my father had known more about the guerrilla movement massing in the mountains of Venezuela in emulation of Fidel Castro's Twenty-Sixth of July Movement in Cuba, and if he had understood the strong anti-American sentiment among university students as U.S. intervention in Vietnam escalated, he never would have let me set off for Caracas. Fortunately he remained unaware, at least until the spring of 1965, when U.S. troops invaded the Dominican Republic to prevent the return of Juan Bosch, the elected social democratic president. That disastrous intervention provoked a civil war that lasted six months and fueled anti-American demonstrations across Latin America. Especially after an American Peace Corps volunteer was accidentally killed on the streets of Caracas, Dad became alarmed. He agreed that "what we are doing in the Dominican Republic is highly immoral," but he worried about my safety and warned me sternly against participating in any demonstrations. His Freudian convictions emerged full force. "Such is the bestial state of man that he cannot think of solutions short of hatred and assault," he wrote, asking me to excuse him for acting fatherly and advising me to hold my tongue and wait for a better climate to express my feelings. "The savage, paranoic fear and selfish possessiveness and jealousy of men make them prey to the worst aspects of infantile fixations and their consequences." He urged me to be careful, because he did not want me to be "a martyr for any cause, especially for one which betokens idiocy on both sides."[4]

I did not agree that morality was "gone on both sides." And in reality neither did he, since he had just condemned the U.S. action. But Dad at this time—and at various times in the future—took a dim view of politics and politicians across the board. He had been reading Nietzsche and felt sympathetic to Nietzsche's "feeling of repulsion for all political movements; to States in fact." In moments of anxiety my father could sound more like the arch conservative Ayn Rand than like the anarchist he imagined himself to be, as when he dismissed "the Panthers, the Communists, the liberals, the conservatives, and the whole kit and kaboodle of diddlers." Dad believed he was being radical when he wrote that "perhaps the only goal is individual freedom and elevation. Whenever systems are invented to improve the human race they turn out to be prisons."[5] But once, years later, when I called him an arch conservative, he did not like it. So was Freud an arch conservative because he was pessimistic about humankind, whereas Einstein was not because he was optimistic, Dad retorted. No, Dad thought Freud was radical, and he did not like being called an arch conservative.[6] And in fact he voted for Ralph Nader in the 2000 election rather than Al Gore, claiming that such a protest vote could show whoever won office that they were beholden to swing voters like him and that even Democrats had to stand up to "the oligarchs who now run our country."[7]

In addition to reflections and news of his travels, Dad's letters over the next several years were filled with paternal advice and encouragement, especially in relation to graduate school, future careers, our family relations, and, of course, love. Sometimes he was practical and specific, urging me to write an article about Caracas, get some tough criticism, and submit for publication. "University people adore publication," he wrote when I was in Venezuela. "With even one small thin contribution and your name in print you'd have an infinitely better chance for many things." He encouraged me to "develop a point of view and hammer away at this point. Interview Venezuelans. Perhaps they can give you all you need for an article; say some subjects such as A Venezuelan Looks at America; a Venezuelan Predicts the Future of His Country; Art in Venezuela, etc. I don't mean to arrogate to myself the specifics, merely to suggest a way of approach."[8] He didn't think the field mattered as much as a good approach to the field. That and some visibility through publication, he advised, could make it possible to get grants for research, attend conferences, and teach abroad. "One must not get stuck in some humdrum job no matter what the field," he warned. "The way out is wide open for any energetic attack by any person with a typewriter and an idea." He worried about my lack of focus: "South American affairs, ethnomusicology, sociology, teaching, filmmaking, etc." He urged me to decide on a subject and stay with it; as long as they did not teach one class after another "in stereotyped fashion," he thought academics could lead interesting lives.[9]

I did not understand until later that Dad's wise advice and encouragement of a profession for his daughters was not the norm among middle-class American fathers in the mid-1960s. While nearly half of all American women worked outside the home, most were in the so-called female jobs of elementary and secondary school teaching, nursing, clerical work, librarianship, and domestic labor. The percentage of women in the professions, including psychoanalysis, had declined since about 1930, beginning to increase significantly only in the 1960s. While Dad could hardly be called a feminist, he strongly supported his daughters' aspirations toward higher education and any profession in which we were interested (though he discouraged my attraction to journalism). And there was always the shadow of Huntington's complicating all thoughts of marriage and children. He wanted to make sure we could support ourselves on our own.[10]

In his letters to Nancy and me, Dad returned often to the idea that it was better to take risks and make mistakes in life than to stand still out of fear and indecision. In a letter to me, he wrote,

> Being at the crossroads is more characteristic of life than being settled in one's
> aims. Perhaps that is the normal situation, being at the crossroads. Maybe we are

constantly at that point because every road leads to some crossroad. Maybe that is the nature of things and that we must always question, no matter what road we have elected to travel for the time being. The only thing that I can add to this is that it may be important to make some election without ever being certain of the validity of the election. Maybe we have to commit ourselves in one direction for the time being, knowing full well that at the next crossroad we may take off in some different direction. I think the essence of the problem lies in the failure of some people ever to make a commitment for fear that they may find a different alternative later which will be more attractive. This leads to something so static and indecisive that nothing is accomplished. One should not object to being at the crossroads but only that one stands still there. Just sum up what you can of the situation as carefully as you can, commit, and then be prepared to re-commit your forces in other ways if it doesn't turn out.[11]

Dad always encouraged us to do something creative, to discover a passion and pursue it. He understood, certainly more than I did at the time, how fortunate we were to have the luxury of choice, and he wanted us to take advantage of it. "Don't spend too much of your life doing what you don't want to do," he warned. "That's sadder than any lost lover. Just a lost life."[12]

He responded to my many complaints of love troubles with a blend of tenderness and toughness. Other young women might have shared these heartaches with their mothers, but since our mother had long ago abdicated the role of adviser, Nancy and I hardly ever confided in her in this way. Besides, I thought, counseling was his profession, his expertise. Why shouldn't I take advantage of that? "Don't play for small stakes, don't do what so many do—refuse to give up the unsatisfactory but safe until something better turns up," he advised. "And don't play off one man against another; or hang on to one because nothing is presently in sight or hand. From my own experience that way lies disaster. Each situation should be judged on its own merits. Yes or no. Otherwise your life amounts to constant compromise." He knew of what he spoke, I felt certain. "And never, but never, be afraid that no one will ever show up. . . . Someday you'll either hit—or you'll not. Big joy or big misery. But any misery of that type is always tempered by two things: self-respect for not having drifted into a morass of half-assed settlements for less; plus the bigness of the misery which, to my mind, is easier to deal with than the slow desiccation of all ideals, all hopes, all aims."

That was his personal value system, he wrote, and he urged me to choose my own. "As long as we respect each other we can never be totally wiped out."

He wanted us to remember that "a truly creative, gratifying, synchronized, and satisfying relationship or relationships are hard to come by." Just because experiences seem to end up with dissatisfaction, rejection, or loss of love does not mean necessarily that one is doing something "cockeyed," as he put it. It is also part of the human condition.[13]

In one of his most eloquent letters, in 1967, he laid out much of his philosophy of life with a deep sense of humility and compassion. "So far as I know, there are no answers," he wrote in reply to a letter I had sent about a boyfriend who had begun seeing another woman. "Only opinions, and yours is as good as mine. But at least I can offer an opinion. Or opinions since there seem to be a number of painful questions." One of my questions concerned infidelity, which must have touched on a sore spot with him. "I doubt very much that it is possible to have any long-term, meaningful relations between men and women on the basis of total freedom," he wrote. "This implies the absence of commitment and without that the relationship is necessarily thin and without too much substance." Of course one may prefer it like that, even if, in his view, it is "missing the mark": "It is really a matter of what one wants; sometimes a matter of what one can tolerate. I hate to put it down simply to a matter of taste or even character but to a large extent that is what it is." He believed that it was better, while one was young, at least, "to shoot for the moon. That is to try to make a deep and abiding relationship on the basis of real love and real commitment. For me and I guess for most people that is the very best. It is the most rewarding potentially even though the most difficult; it is also the most painful if it doesn't work out." He reiterated his view that "big gains are obtained only by big risks."[14]

Dad also addressed the issue of jealousy, which he did not think was "altogether pathological even though it may become excessive. But there is a normal jealousy and one shouldn't be ashamed of it." About loneliness he wrote, "I think most of us feel pretty much alone. It is damnably hard to be understood by another." He thought that he, Nancy, and I did better than most and that "all one can do is to develop the kinds of tastes and interests that one believes in or enjoys . . . and then see if you can uncover another person who swings to the same tune; or develop someone who learns to do it." But that was just a way of reducing the general state of aloneness that we all experience to some extent, he added. "And even that state has its own compensations because it can be seen in some lights as the heroic struggle we all go through to overcome the barriers erected by our skins and minds."[15]

The value of struggle was a theme that wove through much of Dad's correspondence during this time, no doubt reflecting his ongoing efforts to craft a

fulfilling life for himself but also, as always, his anxiety about Huntington's—
never far from his mind, as I have learned from his letters to Henry. He was
extraordinarily tough minded and unsentimental, even before Mom's Huntington's diagnosis. He thought good friends were about "as much as we can demand
in life which seems always to be a series of disasters with interludes of surcease
and pleasure." Was this the therapist speaking, one who saw people at precisely
their most vulnerable moments, often in the midst of a crisis and in pain? He
did not think he was being totally pessimistic. He thought of life rather as "a
gallant struggle against great odds and if you come out on top or near enough
the sun to get a good psychic tan it is more than enough." Speaking now directly
of his own life, he wrote, "I've long ago given up thinking in terms of ultimate
victories, ultimate accomplishments, ultimate relationships and learned rather
to expect and even enjoy the process of overcoming difficulties and leading a
rather day-to-day life of matching my resources against what life has to offer and
signing off each day with some real satisfaction if I've given it the best I have." He
admitted it wasn't much by way of "profound wisdom or lightning bolt insight,"
but for the moment it was the way he felt rather than thought, and he hoped it
would approximate an answer.[16]

He was unendingly supportive, almost too much so. Once, when I was robbed
in the street in Caracas and told Maryline but not him and he discovered it by
accident, he wrote, "Don't worry about my worrying. Also don't think I'm not
able to take your troubles and mishaps. No matter what happens in any sphere
and without reference to fault, negligence, etc., you ought to know I'm on your
side and will back you up."[17]

Though I have few memories of my father training his psychoanalytic scrutiny
on his daughters, he did venture into this terrain with me in the mid-1960s
as he became concerned about my continuing career indecision at the age of
twenty-three. This indecision was possibly "an unconscious effort to avoid meeting the ultimate fate of growing up and assuming responsibility. You should
think about it carefully." When Nancy was feeling insecure about her intellectual
competence at Radcliffe, he wrote that her anxiety seemed to stem from some
"notion that there are secret elements that really professional people have about
research that make them possessed of secret and special tools that you don't have.
Mostly nonsense," he assured her. "Don't be so prepared to wipe yourself out as
helpless, hopeless." Above all, "don't get tied down to being a child looking for a
father advisor."[18]

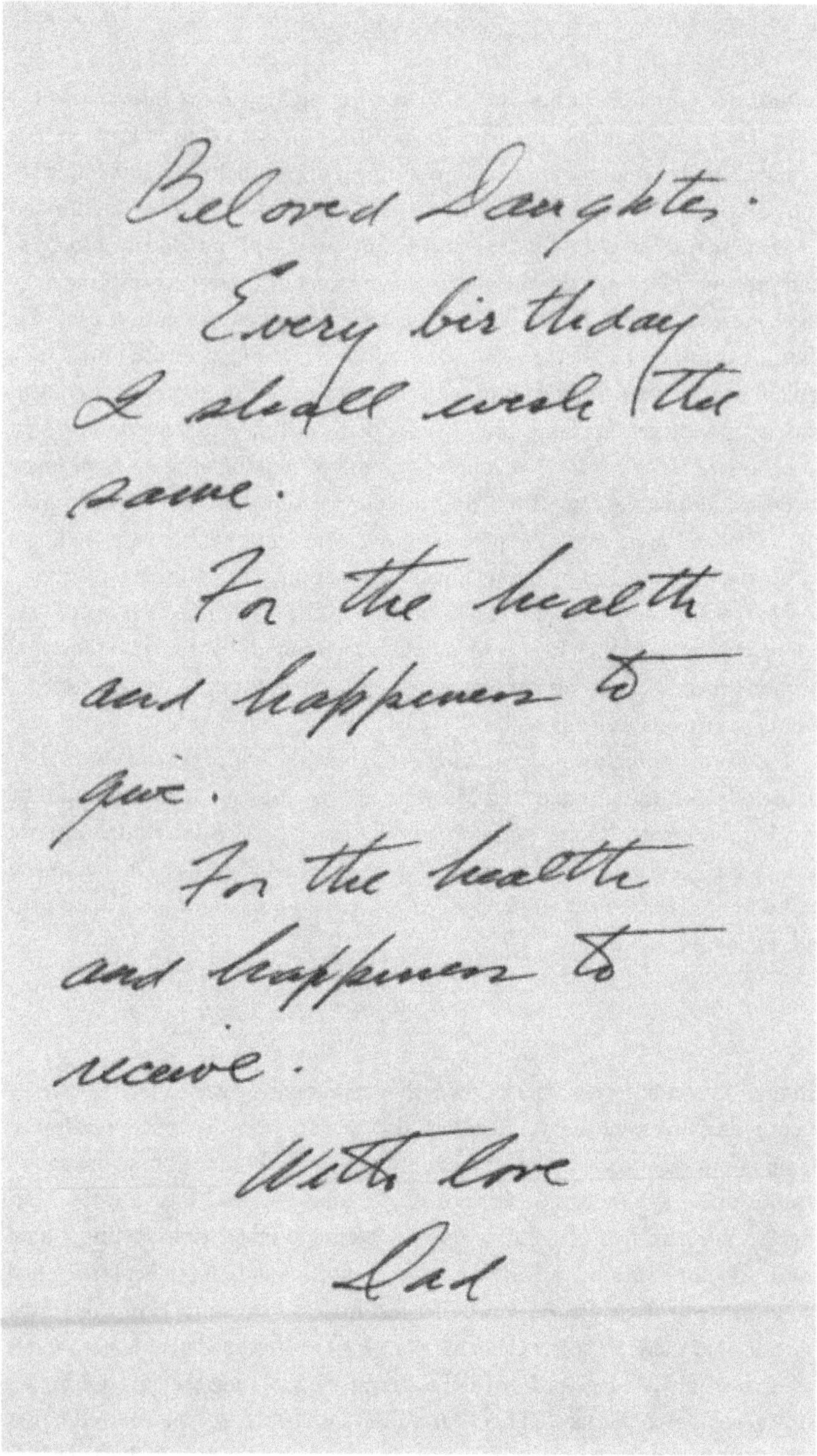

Beloved Daughter:

Every birthday I shall wish the same.

For the health and happiness to give.

For the health and happiness to receive.

With love

Dad

FIGURE 9.1 A birthday card from Dad to me, circa 1970s.

Dad's most explicitly psychoanalytic advice had to do with what he felt was "my overestimation or idealization of his views." The theme emerged in full force in early 1965, when I was still in Caracas, and it continued for several years. "I'm not about to give you advice on your life and loves," he wrote in January 1965. "Not until you get me down to some kind of level like any other ordinary human being, whose dictates you needn't follow, or even be influenced by." I needed to realize that he had his own uncertainties and "philosophical and life difficulties" and that his comments to me might have value not as the dictates of a father but as expressions of his struggling efforts to understand and communicate. He wondered why he was such a failure at making me feel at ease and accepted just as I was. "I can't seem to mention certain subjects to you or to Nan for that matter without it seeming to both of you it is a criticism," he wrote. In his view, our family problem was not that we didn't know each other; it was that what we often imagined was there—such as negative judgments or criticism on his part—really wasn't: "I guess it isn't too far from the truth that we often tend to worry each other and for no real causes but because of some fantasies we hold about each other."[19]

At one point Dad became so wary of offering me advice that he suggested I write (again) to Romi, a disastrous idea, as he later acknowledged. I should have known better, considering Romi's previous letter to me. His reply could have come straight from Helene Deutsch or Marynia Farnham. "If someday you want to get married and have three kids and cook and entertain and decorate and help your husband," Romi wrote, "you ought to be someplace where you will be accessible to such a man. I do not think that traveling in Latin America offers much hope for that." To attract a man, a woman had to be glamorous, sexy, or financially independent. And since, in Romi's words, I had "never been one to emphasize glamour," I should become financially independent, especially since he felt American men were insecure enough that they needed women who were able to support themselves. Romi recommended teaching, that is primary or secondary school teaching, which he implied would be less likely than a higher-status profession to interfere with love and childbearing. He did not want to "belittle academic achievement or development of the mind or potential," but he believed all this could be done while still keeping "in mind the objective of keeping yourself accessible to the greatest exposure to potential lovers and husbands." In his view, teaching was "a form of loving, and if you like people and above all like mothering, then teaching is a great preparation and also a substitute."[20]

I was livid! Even then, only one year past Betty Friedan's *The Feminine Mystique* and with second wave feminism just beginning to gather steam, I found Romi's reply insulting and demeaning, with its entangled racist and sexist logic.

Dad, too, was taken aback but allowed that perhaps Romi's retrograde ideas upset me only because I tended to look on them "as if they were directives":

> They are, after all, only the directions in which we look, perhaps not at all accept-able to you, but then perhaps they may turn out to be the starting point for your own associations and creative thinking. If they are cockeyed then put them aside and mark us both down as duds in terms of adding to your own thinking. It is your life and you must live it as you see fit. But you must add to everything else that sense of toughness which allows you to look at all sides of the coin and take from it the pictures that please you.[21]

It was as if Dad kept forgetting that he and Romi were revered psychoanalysts whose words carried immense authority, and not just for me. One daughter of a psychoanalyst put her finger on the problem: it was difficult to ignore our par-ents' advice because they were the experts.[22] It seemed unfair for Dad to imagine that I could easily laugh off counsel from esteemed psychoanalysts just because they happened to be my father and his best friend. While by the mid-sixties clas-sical psychoanalysis was under assault from many directions, analysts such as my father, whose practice had turned more toward face-to-face psychotherapy, remained admired and sought-after figures. Other people, too, found him formi-dable, and that was comforting to know at a time when I was feeling uncertain and insecure. Dad even told me about people who let him know that it was easy for them to put him on a pedestal. "And then," he wrote, "because they have ele-vated me so much, want so much, need so much, [they] tend to see my every word as a challenge, as fear provoking." He wasn't boasting; he was genuinely puzzled, perhaps because, as a former debater and attorney, he was used to standing up to views he disagreed with and did not find them threatening or even unsettling. The problem, he suggested, was not that he was so strong and wise but that oth-ers needed to have gods.[23] He thought I should consider the possibility that it wasn't so much that he was dominating as that I lived in anxiety and timidity. He reminded me that he had once told me that I was afraid of him, that seeing him alone "was sexually and intellectually threatening" and that I had agreed.[24]

I knew Dad was trying to be reassuring by urging me to have more confidence in myself. He was doing what fathers are supposed to do and too often do not do. He kept telling me that he was happy with me just the way I was, that he wanted me to be more comfortable in my own skin. That he accepted me, was proud of me. Eventually he concluded that I was a late bloomer and said he was not going to worry too much because he had been a late bloomer, too. Also he had taken many people off their pedestals, as he also hoped I would do. He wrote,

"The only important thing that you haven't achieved as yet is an acceptance, a joy in your own unique lifestyle that would enable you to walk through life as if you had reached some settlement with envy and fear of rejection." He closed one especially analytical letter with a thought that he would repeat throughout his life: "Being your own person has a deep meaning. Search for that. I think you will find it and get some joy from it."[25]

The years 1964 to 1968 were perhaps the freest of my father's life, despite his increased financial commitments entailed by the divorce. His letters sounded positively exuberant. He had been "swinging around this country of ours at a great rate recently," he wrote to Nancy and me in December 1964, "and so haven't had time to do anything but prepare for conferences, prepare for lectures, and hope for a little sleep between times." He would "have the chance to do some interesting research soon," he wrote a few weeks later, "but I'll know better in a month or so and will then write about it. On schizophrenia of course. And techniques of treatment." "Hi Daughterkins," he responded to one of my letters. He was "sailing off to Boulder, Colorado, for the weekend and a conference on training psychologists. I'll be snowed all of next week since I'm leaving again next Friday for Washington, NY, Boston."[26] Tensions continued within the American Psychological Association (APA) between the academic psychologists (the "scientists"), wedded to laboratory research and quantitative methods, and the clinicians (the "professionals"), who worked mostly outside academia treating patients. Dad was involved in ongoing efforts to broker a relationship acceptable to both, as well as on an APA committee that worked successfully to get insurance compensation for psychotherapy. He also participated in efforts to imagine an ideal training program for psychotherapists. The extent of his involvement in such activities amazed even him. At one point he told his colleague Robert Holt that he was "so astonished to discover myself as a politico that I look in the mirror a little more often these days to see if lines of corruption are beginning to appear."[27]

On the home front there was more. Another letter noted that Romi had found office space in a new building and wanted Dad and Maryline to join him there: there was room for the Foundation for Research in Psychoanalysis which they were forming. "It would be great fun," he told Nancy and me. "The view is magnificent—the mountains on one side and the ocean on the other. There are many complex things to work out but at the moment the dream is exciting." He was reading Jean Genet's *Our Lady of the Flowers*, Kōbō Abe's *The Woman in the Dunes*, and the letters between Henry Miller and Lawrence Durrell, the latter of whom he greatly preferred. He was finding Hermann Hesse's *Magister Ludi* "a gem." He was going to Palo Alto for a meeting in March, the following weekend to New Orleans. He was "saving my pennies these days since I have some vague dream of going to the psychoanalytic convention in Amsterdam in July. Just as an excuse to take another junket through a few more art museums and countries."[28]

And then Robert Hutchins, a former president and chancellor of the University of Chicago, had invited him to spend a year with the Fund for the Republic, a liberal think tank in Santa Barbara, "but I can't afford the luxury of a pure think job, though I would love it." He was off to a conference with Nathan Leites, a professor of politics at the University of Michigan. In November 1965, "I've a dozen more letters to get off which shoots Sunday down the drain. And sixteen books to be read, research to summarize, projects to write up, and there goes the French again. Wow!"[29]

Dad the debater had not disappeared. He was arguing now with Abraham Maslow, an influential psychologist at Brandeis University who would be elected president of the American Psychological Association in 1968, about values in science, and Maslow wanted to publish their exchange. Their ideas overlapped at many points despite Maslow's use of a different vocabulary. Maslow had just sent Dad his new book, *The Psychology of Science: A Reconnaissance*, to which my father responded enthusiastically and which prompted an apology for what Dad regretted had been "an altogether inappropriate contradiction of your views" at a recent APA meeting in Washington, DC. In fact, the two friends agreed that scientists and science were and should be informed by values, a major thesis of the book. Dad wrote to Maslow that he had always acknowledged operating within his own values but that he had made an effort "to hold them in awareness, to act with some humility and doubt." He encouraged patients to develop a similar awareness and tried to "keep in check any temptation to equate my own values with any scientific veridity [*sic*]." Where he dissented with Maslow was in the idea that science could discover such ultimate values as truth, beauty, or goodness beyond what was "presently socially useful, individually gratifying, or generally acceptable." "I don't think you would question that what suits one age or one culture, however scientifically validated, might be totally destructive or repulsive in another age or culture," he wrote to his friend. "Where, then, is this 'scientific-divine' absolute that you suggest in so many places?" Dad confessed that "with age [he was fifty-eight] my skepticism has increased. It should be the other way around. By now I ought to feel I know something absolute. Your 'faith' is admirable; but I only go along so far, then I have to stop."[30]

Besides such stimulating intellectual exchange, there was social activism and art. He was marching in an anti-Vietnam war demonstration. John Altoon was having "a monstro show in San Francisco" with two drawings from Dad's collection included. He was thinking of attending the opening. John and Babs, who had recently married, were going to Europe, and there was a possibility of meeting them there. He was participating with many others "on a big plan to start a Foundation for the Generation of Art." It was not to be a museum or gallery but "rather an eating-meeting place, a library, seminar rooms, etc. for artists and public." The idea was to create spaces for the creation of art, performances, exhibitions, and collaborative efforts. The painter Sam Francis, the art critic Jules Langsner, the curator Walter Hopps, and the composer Karlheinz Stockhausen were getting involved. "So we have some rich guys interested and will see if we can get it off the ground. Will be difficult," he wrote, "but worth a try." The New Arts Society, as it came to be called, aimed "to create a sense of community among LA artists outside the museum framework and foster work that would help reframe and restructure society." Luminaries such as Susan Sontag, Ray Bradbury, and Mark Rothko would comprise the advisory council while the arts council drew on local art stars such as Ed Ruscha, Ed Moses, and Larry Bell, all of whom Dad knew. Eventually the New Arts Society collapsed for lack of funding but according to writer Gabriel Selz, it had "a lasting and profound impact" by planting the seeds for a museum of contemporary art in Los Angeles, that is, MOCA. Besides this immersion in the arts, he was "working like mad on Failures in Psychoanalysis [a research project he was conducting with Romi, interviewing other analysts although the project remained unfinished]. Also with patients. And occasionally—but oh so occasionally—[having] a little fun."[31]

A therapist's practice is confidential and usually inaccessible to the eyes of others; I could read my father's papers and talks, but I could not visit his office and watch him in action. However, some of the letters saved in his archives offer glimpses into his work. And a few patients told stories of their experiences with him that give an inkling of his style: a blend of empathy, insight, humor, toughness, and practical support when needed. A painter described how he had been haunted all his life by a childhood memory of, as he put it, "shitting in my pants in school" and being so embarrassed that he had rushed out of the classroom mortified; he subsequently thought of himself as a shit, too. He said my father made him examine this memory: "He opened it up. I was able to bring it out and work through it," he recalled. He was finally able to let it go. A hulky darkhaired fiction writer recalled telling Dad he thought he was a thin, blonde woman, to which my father replied, "That sounds reasonable to me." This writer summed up the essence of psychoanalysis when he thanked Dad for teaching him "how to think about what I feel and how to feel about what I think." The film producer Lisa Weinstein told me how she had gone to see him after her mother died of a heart attack while Weinstein was on a flight to New York to visit her. Weinstein had been devastated, not only by the loss but also by guilt, as her mother had mentioned some symptoms the week before. Weinstein felt that if only she had gone to New York a few days earlier, as she had originally planned, she could have taken her mother to the doctor and possibly saved her life. She woke up each day crying and continued to cry for days and weeks, unable to do anything else. Finally she went to see my father, crying throughout her sessions. One day he said to her, "Look, you're not God. You couldn't have made a difference. You've just got to stop crying and making everyone around you miserable." And somehow she needed to hear that, she told me, because she did stop. And gradually found her way through her grief.[32]

But he took the opposite approach with another patient. One day my father read in the newspaper of a young woman standing in the surf at the beach in Mexico with her two-year-old daughter who had been struck by a sudden wave that knocked the toddler out of her arms and underwater, causing the child to drown. Moved by this tragedy, Dad found the young woman's phone number and called her, offering to see her every day for a year at no charge. And so she saw him, going painfully through scrapbooks and photographs of her daughter with Dad day after day. "I was so lost after she died," Jodie Evans told me. "It saved my life." She established a grief center to help others cope with wrenching loss. Much later she asked him why he had offered to see her that initial year "and he told me his wife had died of HD and that his daughters were at risk and working with me was his therapy for the fears he lives with daily." She felt her time with my father had been transformative. "To be seen in this world is such a rare gift and he had so much patience with humanity," she wrote me. "He wasn't doing therapy as much as healing. Finding what was in the way and helping to dislodge it by witnessing and accepting. He was a safe place to explore [one's] self without judgment but with appropriate questions to help you make new choices." (He also became her labor coach when she remarried and had another baby.) Jodie had always been an activist and a "giver," involved in California Democratic party politics; her experience with Dad deepened her desire to "give back," to him as well as to others. She worked for a decade as executive director of the organization he established, the Hereditary Disease Foundation. In 2002 she helped found the peace organization CODEPINK in protest against the war in Iraq.[33]

They were not all as accomplished or creative. "I'm not a celebrity, doctor, scientist, or VIP of any sort," one young woman wrote to Nancy and me after Dad died. "I'm just an ordinary person. Yet Dr. Wexler treated me as though I was just as worthy as anyone else. Despite his famous clientele, he made me feel I was worth believing in and he was generous with his advice and encouragement." Two years after her mother died, this woman had embarked on a solo backpacking trip around the world, and Dad offered to write to her during her travels. She would write ahead, telling him of her next destination, and once she arrived she would pick up her mail at the American Express office: "At every location on my trip, there was a letter waiting for me from Dr. Wexler. He wrote more than any other friend or relative. And as usual, his letters were full of encouragement and praise for what I was accomplishing. I have kept all his letters and, as you can imagine, they are quite a treasure to have."[34]

Dad and I were getting along better. When a professional meeting took him to Washington, DC, where I had finally decided to go after the Fulbright to earn a master's degree in Latin American studies (at Georgetown University), I had such a fine time with him that I boasted in my diary about how much I adored his company and how proud I was that he was my father.[35]

And then there were more summer vacations in Europe. Dad was so thrilled with his 1964 trip that he returned for one month each summer for the next three years, combining his travels with a psychoanalytic conference or lecture. Although he traveled alone, he always managed to meet people along the way. He did not tell us immediately after that first trip that, in a small museum in Rome, he had met an adventurous young woman about my age who was recovering from a broken love affair and from the loss of her front teeth in a bicycle accident. She turned out to be a mathematics professor at the University of Amsterdam. So in 1965, when he again traveled to Europe to attend a congress of the International Psychoanalytical Association in Amsterdam, he looked her up. She showed him around the university, enchanted him by taking him to her "Writers' Bar," and revealed "a side of Amsterdam I would never have seen on my own."[36]

I do not know if she and my father had a romantic fling that summer. But a few years later he did have a romance with a young Belgian lawyer whose U.S. identification card turned up in Dad's papers with his address on it. She visited Dad in Los Angeles at least once, probably more than once, and he visited her in Europe. He worried over how his daughters might feel about his relationship with a much younger woman and wrote to us poignantly of his own mixed feelings, describing his sense that a great romance with "a twenty-five-year-old girl sounds too silly and something jokes are made of. Even I think so." Yet he clearly appreciated what he described as the Belgian woman's playful, imaginative, loving character and her pleasure in taking care of him, "all kinds of nice little things. I've missed that for a long time, being my own caretaker for the past many years." And though he could get that "in many places," not in such "an attractive package," as he put it. "Or such an intelligent package. And maybe I should add, in such a decent package." But in the end, nothing came of her plan to move to the United States, and I never was able to meet her.[37]

In Europe, psychoanalysis supplied him with connections. He spent time with Max Schur, Freud's physician, whom he had met in 1950 and who had been Romi Greenson's analyst and later friend. He also saw the larger-than-life Masud Khan, an aristocratic controversial Pakistani-born psychoanalyst living in London who shocked Dad at a dinner party when he revealed an unexpected streak of anti-Semitism along with his charisma and brilliance. Dad went outside the psychoanalytic world to seek other experiences as well. The summer of 1966 found him enrolled as a student at the Alliance Française in Paris, although an artist friend whom he had met in Venice the previous year, the "great enthusiast" Jean-Jacques Lebel, an organizer of the first European Happenings, discouraged him from doing so ("For what? Nothing much to read in French nowadays").[38] Dad's Parisian photo ID card shows him with an enigmatic Mona Lisa smile and his shortest crewcut ever.

Of all Dad's European trips during the sixties, that of the summer of 1967 was the most memorable, largely because, by "a fabulous break," he ended up on a cruise ship from the Italian mainland to Sicily with a Brown University art history professor named Bates Lowry who had just been appointed director of the Museum of Modern Art in New York. Dad got to know Lowry and his family aboard the ship and then traveled with them through Sicily in an impromptu car caravan, later meeting up with them in Florence and accompanying Lowry to sites of art reconstruction. "More fascinating than any detective story," Dad reported to Nancy and me. "The scientific and art knowledge that goes on here is out of [this] world!" On the same trip, he took a cruise around the Greek islands, striking up a shipboard friendship with the associate Supreme Court justice Hugo Black. He said that Justice Black "is all for showing anything, painting anything, and if you think it is pornographic, don't look! For an eighty-one-year-old ex-Alabaman, ex-Klu [sic] Kluxer now reformed, that is quite something. He is one of the great liberals of the court now and quite a man." On his way from Greece to Copenhagen for meetings of the International Psychoanalytical

FIGURE 9.2 Studying French in Paris, summer 1966.

Association, he stopped off in Saint-Tropez, where Lebel was putting on Picasso's Surrealist extravaganza, *Desire Caught by the Tail*, under an enormous circus tent on the beach, with such underground performers as Taylor Mead and Ultra Violet of Andy Warhol fame and music by the English rock group Soft Machine.[39] I doubt my father knew who anyone was or of the illustrious context of the play—Picasso wrote it in 1941, and Albert Camus directed a reading of it in 1944 in the Paris apartment of Michel Leiris, with Simone de Beauvoir, Jean-Paul Sartre, and Picasso himself among those reading parts. And he never found

Lebel amidst the crowds. But he had a "wild time . . . with the craziest hippies and intellectuals I ever met," he told us later. After such adventures he found the Copenhagen meeting "dismal." "God how stuffy analysts are!" he wrote. "So I got me a bike, hung out with my Mexican friends, met some really pleasant Argentinians, and had me my own fun. It was nicer riding my bike to the zoo than any meeting in the land."[40]

By the time of the disappointing Copenhagen meeting, Dad had grown thoroughly disillusioned with his colleagues in LAPSI, owing in part to the LAPSI board's disastrous handling of a complaint brought against him by a female former patient. At the time she was no longer seeing my father, who had terminated her therapy two years earlier on the grounds that they were not making progress and she would do better with another therapist—a decision she resisted, according to Dad's later account. In July 1964, this patient made a complaint to the LAPSI president, Leo Rangell, who notified the board of directors. But it was not until December 30, six months later, that one of the board members, Maurice Walsh, called to notify Dad that a complaint had been made, demanding that he meet with the board to discuss it. When my father asked the nature of the complaint, he was told that he would be informed at the meeting. As a former lawyer who understood due process, he refused to attend the meeting without hearing the complaint beforehand. He was further incensed when he learned that the complaint had been lodged months earlier without anyone letting him know and that board members—constituting themselves as a grievance committee— had been meeting about the issue since that time. A standoff ensued, leaving the right of the patient to have her complaint heard in a timely fashion ignored and my father accused of something about which he was uninformed and without a fair hearing.[41]

This Kafkaesque situation dragged on for several more months until, in rage and frustration, at a meeting of the LAPSI membership my father demanded an investigation of the board, accusing them of acting in a highly prejudicial way, motivated "by personal animus, malice, and hostility" and using procedures that were "undemocratic and reprehensible."[42] It escaped no one's notice at the time that my father was closely allied, both personally and professionally, with Romi, a critic and rival of Rangell, who had overseen the investigation for the first six months.

Nancy and I learned something of this battle in the spring of 1965 when Dad mentioned to us that he was "fighting a whole bunch of stinkers in the Society and at the moment we are on the verge of a real climax that may blow the whole Society to smithereens. Either I'll get killed or they will and I think it will be them. Anyway I've got them rocking at the moment and I've some powerful allies in Romi, Ekstein, et al. So life is not dull and I'm wishing it were." A month later he was almost exultant. "I had to get up and make a speech against all the officers and members of the Board of Directors and move that they all be investigated for their nasty tactics against me," he wrote in May 1965. "My motion was carried almost unanimously but believe me it was a bombshell in that Society. It's a big step and even a big victory for me but the toughest part is yet to come. Don't know where all this is going for me but it is unpleasant and hard work and I wish it were over. Like wasting good time and energy on pure crap." This was our father the lawyer and debater, almost relishing the battle.[43]

Within a few weeks, though, Dad's fighting spirit had turned to disgust. He had hoped that the five members of the ad hoc committee established to investigate his charges would censor the board for its shabby handling of the issue. Instead they defended the board's procedures as entirely legal and free from personal animus or prejudice. They even praised board members for their dedication and sense of responsibility. While "judgmental errors in procedure did occur," these were a result of defects in the bylaws, which needed revision. Most infuriating to Dad, the ad hoc committee (which included three of his supposed allies) recommended that the LAPSI membership as a whole give a vote of confidence to the board of directors for their "good faith and integrity" and censored my father for "the intemperate nature" of his response, which they found was unjustified in the situation. According to his account many years later, my father was so infuriated that he got up and walked out, resolving never to return.[44] By this time, however, the LAPSI board of directors had dropped the patient's case entirely, leaving her completely—and scandalously—forgotten.[45]

"Some day I'll fill you in on the details," Dad wrote to Nancy and me, "but at the moment I've had a belly full and don't even care to think about it." He was "kind of through with that group."[46] Dad's fury at the procedures of the LAPSI board extended to his analytic colleagues more broadly. To his brother Henry, Dad let loose with his anger:

Why doesn't anyone do anything? That is the problem of the world today. Because it takes effort, courage, involvement. Who can be bothered? It's easier to throw it in the waste basket. Like we all do about the great issues, the human miseries. Do you think analysts are any different? Yes, they are. Much worse! There will be no rebellion, there will be no change. These people are dead in a filthy, narrow, destructive way. They are mainly zeroed in on the cash box, status, recognition, and keeping out of trouble. They are the Establishment and have an investment in keeping the boat from rocking. They want peace and quiet and they analyze that way too. They are jerking off under the pontifical robes.

Dad felt bitter and hurt by those who had failed to come to his defense and now wanted to be friendly with him. He even felt let down by Romi, who "acted with inordinate stupidity—or shittiness—in my affair of the Society," although Dad insisted that he did not hold it against Romi, because "he is an actor, needing an audience, the spotlight. He would be lost without the group."[47]

Whatever his disgust at that moment with many of his analytic colleagues in Los Angeles, my father was full of enthusiasm on other fronts, still immersed in psychoanalytic lectures around the country, seminars, conferences, continuing discussions on schizophrenia, planning for the new arts center he had told us about, and studying French. "I'm deeply involved in all kinds of things," he wrote to Henry in March 1967. He was loving Susan Sontag's new book, *Against Interpretation*. He was discovering recently translated writers of the Latin American literary "boom," including Jorge Luis Borges and Gabriel García Márquez. It was he, not my professors, who introduced me to the nineteenth-century Brazilian writer, Joaquim Maria Machado de Assis, and his devastating, heartbreaking, and funny novels and short stories.

Dad's exuberance was contagious. I could almost share it myself.[48]

In January 1968, Dad and I visited Nancy in Kingston, Jamaica, where she was spending a year (ultimately half a year) on a Fulbright fellowship, just as I had in Venezuela three years earlier. We had a great time together, so much so that I boasted in my diary about how "Nan, Dad, and I really have a good relation[-ship] and it's something that means a lot to all of us." I even admired the camaraderie of Maryline and Dad. In any case I was thinking less about home than my unraveling love affair with a fellow graduate student back in Bloomington, where I was finally pursuing a PhD in history at Indiana University. I knew that Dad had been moving into a different world, spending time with the artists he admired and enjoyed. "What does your father do with himself?" Romi asked me plaintively when I visited the Greensons on a home visit. Apparently Dad never saw them anymore. "He's a strange man, your father," Romi told me. "I love him.

I admire him. I respect him. But I'll never understand him." I did not tell Romi that Dad had new friends and new interests, that he felt Romi did not listen to him. That he had moved on.[49]

In July, Dad wrote to Nancy and me about our upcoming reunion in Los Angeles, where we would celebrate his sixtieth birthday and then take a leisurely road trip across the country, stopping in Yellowstone National Park on our way back to our respective schools, me to Bloomington and Nancy to Ann Arbor, where she was about to begin a graduate program in clinical psychology at the University of Michigan. Dad's letters to us at that time were cheerful and animated. He made no mention of another set of letters to his brother, anguished letters that had been flying back and forth across the country since early May. He spoke not a word to us of the anxious meetings, phone calls, and consultations with doctors that had been crowding his days and nights.

He closed his July letter gaily with the lines "Have fun in France Nan; and Bloomington Ali. And soon we'll have fun all over the U.S."[50]

10

(A) Challenging Fate

Men make their own history, but they do not make it just as they please; they do not make it under circumstances chosen by themselves, but under circumstances directly encountered, given and transmitted from the past. The tradition of all the dead generations weighs like a nightmare on the brain of the living.

—Karl Marx, *The Eighteenth Brumaire of Louis Bonaparte*

om was fifty-three years old and Dad not quite sixty when he wrote to his brother the words he had hoped never to utter, that Mom had been diagnosed with Huntington's chorea, now becoming known as Huntington's disease.[1] He did not keep a copy of this letter, but Henry kept the original in his journal, and his son, my cousin Eric, preserved it and showed it to me fifty years later. The letter revealed that the story Dad had told us about believing women could not get Huntington's was untrue, that he had known all along and had suffered because of it. Perhaps he came to believe his own story, projecting back into the past a fiction he invented after Mom's diagnosis to explain his inattention to the changes in her behavior. In Mom's immediate family, it was true that no women had gotten it until now, at least none that he knew about. It had been eighteen years since Mom's brothers were diagnosed, eighteen years during which time they had lived three thousand miles away, visited only on occasional trips back east. Jesse, Paul, and Seymour were all dead now, and none had had children. And now the illness had come for Mom. A police officer had called her out one morning in the street for being drunk, a common misperception of people with the stumbling and unsteady gait of Huntington's. Alarmed, she had phoned Dad, who arranged an appointment for her with a neurologist, asking that if he diagnosed Huntington's—which my father immediately suspected—that he withhold this information for the present.

Dad began a frantic search for information, recruiting Henry and Romi to the effort. As MDs, they had medical contacts my clinical psychologist father did not have. The information he received was anything but encouraging, however. "There is nothing approaching a cure or even a satisfactory treatment," wrote Charles Markham, a neurologist at UCLA who was part of the Research Group on Huntington's Chorea, formed the year before under the aegis of the World Federation of Neurology. Markham offered the standard eugenic counsel that "those people who might get the disease (those whose mother or father have had the disease) can avoid having children and can best bring the disease to a halt in one's particular family."[2] Tough advice for a father to have to give to his two daughters in their twenties in the late 1960s—or any time. Even Ntinos Myrianthopoulos, a geneticist at the National Institutes of Health in Bethesda, Maryland, and one of the most compassionate of the early Huntington's researchers, believed that "the most important and pressing problem in Huntington's chorea" was developing a method of prediction, or "early detection" as it was called, that would make possible the "wise counseling of individuals" and "prudent management of their personal lives."[3] Just how traumatic such prudent management might be for young people like Nancy and me was evidently not a consideration.

Faced with this clinical desert, Dad focused on finding out about all the places where relevant research was going on, especially research into the fundamentals of the disorder. He was impressed that Wilson's disease had been brought under control when researchers identified a problem with copper metabolism, and he wanted to know about all such approaches. He doubted scientists could solve the genetic problem of Huntington's anytime soon, but if there were a metabolic disturbance, as in Wilson's, it might be possible to correct it by supplementing a lacking enzyme or eliminating an excess of one. He was interested in the problem of early diagnosis but more in biochemistry and its possibilities for treatment.[4]

Henry immediately took on Dad's dilemma as his own, sending out inquiries to his East Coast medical colleagues, mobilizing Dad's former law partner, Henry Sternberg (and Henry's friend), to find a bibliography on Huntington's at the library of the New York Academy of Medicine and mailing Dad long

Dear Hank,

It will not be news to you I know. Lee has the same thing her brothers had. It has been somewhat slower in coming but will not be less deadly.

It is the great poison in my life. Now I face telling her, and quite soon too. But the nightmare is the children. Do I have to tell you what that has done to my life; and will do in the future? And what it has meant all this time when the children were mentioned?

Alice will be here in mid-June and will undoubtedly see something. Nancy will return in August. For me there is only dread in the air. It will be the same for them.

Do you know of any research? Any articles, papers, workers in the field? I don't look for miracles. Only a shred of hope to offer. Let me hear.

Confidential - please!   Best,

                                        Milt

FIGURE 10.1 Milton to Henry Wexler, May 1968.

Wexler family collection.

letters with references and replies to the many questions he raised. Within a few weeks Dad was in touch with several nodes of Huntington's disease research: besides Markham and John S. Pearson, a clinical psychologist in Kansas, there were John Whittier, a neurologist at what was then the Creedmoor Institute for Psychobiologic Studies (later the Creedmoor Psychiatric Center) in Queen's, New York; André Barbeau, a neurologist at the Institut de recherches cliniques in Montreal; and the Committee to Combat Huntington's Disease (CCHD) in New York, which Dad immediately joined. Marjorie Guthrie, widow of the great singer-songwriter Woody Guthrie, had established CCHD after Woody died with Huntington's the year before. She, too, was a charismatic presence, a former dancer with the Martha Graham Dance Company who opened a dance school on Long Island following her separation from Woody. Despite their divorce and her remarriage, Marjorie oversaw Woody's care starting in 1956, when she stepped back into his life. She wanted to make sure their two children, Nora and Arlo, spent time with their father during his illness, including during the long years when he lived at Greystone Park Psychiatric Hospital in New Jersey, where she saw firsthand how little most doctors knew about Huntington's. CCHD aimed to educate physicians and family members and to lobby in Washington for government-funded research and support services for families. Coming from the cultural left, Marjorie envisioned a grassroots association with regional chapters around the country. She began traveling outside New York—and would eventually travel internationally—seeking out interested physicians and calling on families with Huntington's to come out of the shadows. She would tell family members, you can't do research on ghosts.[5]

In those early grim days, it was Henry most of all who listened to Dad's anguish, pursued information, and tried to offer what comfort he could. Dad found it puzzling that all four Sabin siblings had developed Huntington's whereas in the previous generation, just two of four had. He was looking for a miracle, he told Henry; that is, information about inheritance that would make it plausible that both Nancy and I could escape the disease—and if that was impossible, then odds that would at least allow us to have a fifty–fifty chance of escaping it. "If you must know I have for many years been of the impression that they would have NO CHANCE!" he told Henry. "And that was very hard to live with."[6] Neither Dad nor Henry was inclined to dramatize, but their letters of 1968 and 1969 reveal the depth of my father's grief and guilt, expressing an intensity of emotion he sometimes alluded to but rarely showed, at least not to me.

Dad had been thinking about parents passing on hereditary illnesses to their children as far back as 1951 when Henry Sternberg invited him, as a clinical psychologist, to address the first meeting of the Muscular Dystrophy Association, made up mostly of parents of children with the disease. Though he was speaking as a professional, parental guilt must have been much on his mind at that moment, when Nancy and I were six and nine and he had just learned of the presence of Huntington's in Mom's family. His talk came straight from the heart. Parents often felt "tempted to rage against God, their own parents, and Fate itself for such an unkind blow," Dad noted. "And such rage, in itself, is a source of guilty feelings. Partly because, with any hereditary disease, the parents feel the most immediate sense of responsibility for the disaster visited on the child." He warned parents against ostrich-like avoidance, extreme stoicism, and either pretending things were normal or being overly solicitous and self-sacrificing. Instead he recommended honest, forthright communication and talking about their feelings to help the child find "the inner strength and conviction which will enable him or her to make the choice to live fully and happily within the limitations of his [or her] illness." As if anticipating his own future, he told parents that "it would be of immense psychological and practical profit to themselves as well as to their children if they would devote a substantial portion of their energy and means to the support of this Association, dedicated as it is to the solution of the problem of muscular dystrophy."[7]

Seventeen years later, he must have felt even closer to becoming one of those parents. In the face of Dad's determination to confront the facts head on, no matter how depressing, Henry tried to offer solace, telling his brother, "Don't equate fact with gloom and non-fact with reassurance. If you do, it should tell you something about yourself."[8] Henry offered wise advice that Dad would pass on to others: "Above all, we must not play God. I have seen too many ironic twists of fate follow from that: by it I mean that we cannot predict outcomes, length of or onset of processes, etc. Nor must we decide that the children are inevitably victims of this disaster. IT NEED NOT NECESSARILY BE SO! Nor is this wishful thinking even though my wishes enter into it. It is based on a reading of statistics. Let us not visit the disease on them until it actually is there. Pray God it never is!"

He added that Dad had been "heroic carrying this burden alone. Too much so." He had not allowed himself the comforts of sharing these agonies. "I don't say this by way of censure. You know it as well as I; it is merely to say that I now know it as well."[9]

Henry took Dad's anguish deeply to heart. Like most physicians of his generation, he had minimal exposure to genetics. But he tried to explain to Dad the dominant inheritance pattern of Huntington's and offered whatever biochemical information he could find. Dad was a tough interlocutor, often angry and critical of his brother, although Henry understood well that, as the source of bad news, he was a clear target. It was typical of my father to thank Henry and then immediately criticize his findings. Henry did not seem to mind. Building on his brother's emphasis on fundamentals, Henry said that they must find people who have a "bold and imaginative approach to the disease," such as James Watson, a co-discoverer of the double-helix structure of DNA, Linus Pauling, and Albert Szent-Györgyi (all three Nobel laureates), who might be interested or know others who were.[10]

Within a matter of weeks, the two brothers had laid the conceptual foundation of what would eventually become the Hereditary Disease Foundation. As Dad once advised me, "Seeds grow plants." But perhaps the most telling word in all of Henry's many letters to my father from the summer of 1968 was the shortest: *we*. "If we can prevent some biochemical action," he wrote, and "if we can in some way neutralize it." It was that beautiful plural *we* that jumped out at me in the winter of 2019 and 2020 as I read the correspondence of these two psychoanalyst brothers who closed their letters with a formal "all the best" or just "best" but who relied on each other and opened their hearts to each other as to almost no one else in all the world.

Meanwhile, in his journal, Henry wrote down private thoughts about Dad's situation. "What I suspected when I last saw Lee has come to pass," he wrote on May 18, 1968, shortly after he heard from Dad. "Now the prospect of Alice and Nancy! Why had Milt lied to me when he told me a long time ago that if Lee passed her fiftieth year free from symptoms she'd not develop the ailment?" When Henry asked, Dad replied that he had not told the truth "because he wished to preserve the fiction that all would be well." "What bleak prospect," Henry wrote. "I have a heavy heart." But Henry could not help casting a backward look that reflected the continuing stigma surrounding Huntington's within the medical community: "Why did Milt take this upon himself? I recall warning him when he went with Lee and he didn't heed it. I think I understand the divorce now . . . his wish to separate himself from what he must have sensed was coming on . . . a running away when it was too late. Too late."[11]

These thoughts surprised me when I read them in 2019, for they indicated, in the first place, that Henry apparently did not know the marriage was de facto over long before Mom showed any symptoms. No doubt many of Dad's friends and colleagues also assumed he had wanted to flee future caregiving obligations, a misunderstanding that must have added to his pain. Had they known about Maryline, they would have understood the divorce differently, but of course he was not about to tell them, especially since he and Maryline continued to live separately and appear in public only as colleagues and friends.

More important, Henry's undertone of blame bears further witness to the eugenic thinking that persisted within the medical profession well into the 1970s. Even Dad shared it to a certain extent, for he became angry and defensive when I asked him, more than once, just when he first knew about the presence of Huntington's in Mom's family. He always insisted that he had not known her family history until her brothers were diagnosed in 1950, years after Nancy and I were born. "Do you think I would have had children if I had known?" he would respond angrily. *And what about Nancy and me?* I would think to myself. *Are you telling us that we should not have children?*[12]

Henry assumed that beyond the pain of receiving such a dreaded diagnosis, Mom must have felt "a bitterness of rage" against her parents for bringing her into the world. "Not that we all don't have an inevitability of death," he added, "but to know so much beforehand it[s] inexorability in the fashion it will take!" Henry wondered whether Dad had decided on the divorce himself or had received a suggestion from a physician on what lay ahead and taken steps accordingly. "Was the disease beginning to show itself to him even at that stage and, knowing what was in store, he decided to separate himself from it before anyone could accuse him of desertion in the face of sickness. . . . He certainly went all out to settle his estate on her. . . . Did he know that it would have such a decided foreseeable end point?" Although Henry felt deep empathy for his brother, he alternated among believing that Dad should not have married Mom, that he should have left her sooner, and that he should not have left her at all.[13] His thinking reflected the predominant view in the medical literature at the time: that those with Huntington's or at risk should not have children, that individuals from unaffected families ought to avoid marrying into families with the disease at all costs, and that if those from families with no history of the illness were foolish enough to do so, they were partly to blame and should stick it out, no matter what.

While discussing research, Dad and Henry also discussed at length what to tell Mom and how to tell Nancy and me. They strategized about these disclosures as if they were planning a military campaign. Dad confided to Henry that he was "maneuvering, by lies, mostly, to get both children back here by Aug. 15. I'll have to tell them straight and that will be toughest of all. I dread it but it must be done." To Nancy and me, he merely urged that we come home in time to celebrate his sixtieth birthday on August 24, reminding us of his promise of a road trip. Meanwhile Henry worried that Dad was too emotionally involved to relay such traumatic news to Nancy and me—and Mom—and urged Dad to seek help from others. While Dad allowed that Henry might be right, he was not comfortable with anyone else taking his situation on. Romi had offered to help, but Dad demurred. "He is too self-involved, too sentimental, too prone to tears, but I think they are tears for himself," Dad told Henry.[14] Henry, too, had offered, but Dad thought Henry "too moody [and] depressive" and said to Henry that "what you say may sound portentous."[15] For all his earlier extolling of straightforward and honest communication with respect to muscular dystrophy, Dad wanted to keep the diagnosis of Huntington's from Mom for as long as he could, fearing that the hereditary aspect would be too painful to her. With Nancy and me, however, it was different. He was adamant that we had to know the facts right away. "I'll never allow them to expose themselves or their children to this horror," he wrote to Henry, revealing once again his eugenic convictions. "It will be a nightmare for me and for them. But how else to avoid it? No way out. And again who do you think they will trust and believe most?" Regarding Henry's alarm about some of my father's remarks, Dad replied frankly:

About your alarm, of course I entertained thoughts of ending it for all of us. That is merely among the million things I have thought of, not just now but often in the past. This thing isn't easy to live with. But I am also a rational person, not given to dramatizing myself too much, not given to too much sentimental slobbering, and if I slip now and again I think I'm entitled to it. It's been a long, hard pull and the road ahead is pretty tough. But I don't think I'm apt to act too impulsively. You merely hear now for the first time some of the things I've kept to myself simply because I didn't see any point in getting myself and everyone else in a stew about this when things were more a deep suspicion than a certainty. Time enough to face things when they come along. That time is now and so I share things that otherwise I'd rather handle myself. I'll listen to advice, help, etc. But the final decisions are for me.

Dad disputed his brother's suggestion that he was working too hard and not getting enough rest. "Work is my only psychological salvation," he told Henry. "It stops me from thinking. I'd go mad sitting at home, resting, thinking, vacationing. The only way I got through the past three weeks was to work my tail off, sometimes fourteen hours a day." As for talking to others here, "you can forget that. Romi is trying to be helpful but he is too full of himself." Maryline was "OK but she is that kind of spartan who doesn't really understand this kind of psychic pain; and she loves the kids just about as much as I do. Beyond that I see no one and don't care to. But again work comes to my rescue. And making money and being stupid and dull and yet [feeling] a relief for the guilt seems quite reasonable to me." He allowed that the situation might ease off in the future, "but for now you shouldn't attack a compulsive drive that serves so well as a defense. Agree?"[16]

Dad met Nancy and me at the airport when we arrived back in Los Angeles, by chance on the same day. I came from Bloomington, and Nancy was returning from France, where she had been on holiday with friends, having left Jamaica to spend the last six months of her Fulbright fellowship at Anna Freud's Hampstead Clinic in London. We drove to Dad's apartment in Westwood, where he told us our mother's diagnosis and explained the fifty–fifty risk to ourselves and the risk to any children we might have. Three generations all at once. For ourselves and for future children we might have, we could not know which way the dice were thrown until symptoms appeared. We would have to live with that. But he was going to do everything in his power to mobilize scientists to find a cure.

It was all so abstract, so unfamiliar and strange, that for me, the meaning of Huntington's did not sink in for a long time. Mom seemed only a bit more timorous, anxious, and jittery than when we had last seen her. We spent time at her apartment, affirming at first what Dad had told her, that she had a neurological disorder akin to multiple sclerosis that was causing her unsteadiness. But she clearly knew the truth, which Nancy bravely told her within a week or two. Eventually Nancy, Dad, and I left for our road trip. Dad later wrote about our "marvelous holiday" and how "that trip will feed me for a long time—with all kinds of pleasant memories and verifications and beliefs and expectations," which was not how I remembered the trip at all.[17]

It is tempting for me to see my father's entire life until this time as preparation for the moment he learned of Mom's diagnosis. It was as if he had finally met an opponent worthy of him. He brought with him to the struggle a formidable arsenal of skills and social capital: his experience as a debater and his practice as a lawyer, which had honed his public speaking skills; his marriage to a biology teacher, which had deepened his appreciation for science; his friendship with the father of a young girl with muscular dystrophy—and the girl herself—which had given him insights into the impact of hereditary illness and inventive ways of living with disability; his exposure to progressive social thought and experience doing research at Teachers College; his experience with Nedda, which had increased his confidence in the possibility of intervening effectively even in so-called incurable diseases; his group therapy with artists, which had enhanced his ability to organize creative discussions; his visibility within the arts and entertainment worlds of Los Angeles, which had helped him call attention to Huntington's disease; and his participation in the collaborative, research-oriented family culture of Menninger, which had prepared him to create a foundation family of his own. And most of all his confidence—in the future and in himself—that change was possible and that he could help make it happen. If you had asked him in 1968, he might have said, as he said to me the following year, that he was a pilot often flying by the seat of his pants without a clear plan to guide him, that he was "much more a feeling and less of a thinking person than you think."[18] But he seemed to know, or perhaps to feel, that he had found his mission, the grand purpose that would give shape and meaning to his existence.

Upon returning to Los Angeles, Dad slowly began to figure out how to live now that Huntington's had crashed like an asteroid into all our lives. "However bad the news was, the fact that there were no more secrets had its own therapeutic effects," he wrote later. He could focus on the disease, not on "family complications and dreads." He had a feeling of "cleanliness." "For so many years I had been living with lies, pretense and repression," he acknowledged. "With things out in the open, I had the sense of a restored life."[19] Mom had not collapsed with the diagnosis, but in a short time she became unable to manage on her own, more in a psychological than a physical sense. So in a move unusual for most divorced American men at that time, Dad took over. He did not call us home to care for her as so many daughters have been called, past and present, to care for ailing parents. He wanted us to continue our education and to live our lives with as much freedom as possible. While he did not take his ex-wife into his apartment or do hands-on caregiving for Mom, he found her places to live, hired caregivers, chose her physicians, and oversaw her treatment. He paid all her bills. He visited her most Sundays. "Big circle. All the way from marriage to marriage. Or responsibility to responsibility," he wrote to me ruefully.[20] And unlike the situation for many HD families, he had the capital, social as well as financial, to do all this. "I was able to mobilize many people to help in this crisis," he told Henry, adding that once Mom realized she wasn't alone and that Dad would be there for her, she began to "settle down," as he put it. That helped a lot during the day while he was working. The nights were awful, he admitted, and he was starting to drag.

His letters to Nancy and me and to his brother that fall and winter alluded to his punishing work schedule. "I've stretched my workday a bit, and now I'm up to about a ten-hour stint per day and maybe a little thrown in on Saturdays," he wrote to Henry. "Damned fortunate I'm in demand or I'd be up shit creek." He was suffering from "free time hunger," he told his brother, "in a society that is supposed to be geared to more and more leisure time. Sure am out of whack with that."[21]

Through the fall and winter of 1968 and during all of 1969, Dad felt his way forward, contacting researchers and clinicians and educating himself about the disease. In the days before email and mobile phones, he wrote letters, hundreds of them, to researchers, clinicians, Huntington's family members, and potential donors and supporters, letters he wrote easily and well. He also worked to establish a legal framework for a California chapter of CCHD, which meant negotiating all the bureaucratic stepping stones needed for a nonprofit to operate tax free in the state.[22] Already by December 1968, he had organized the first informal meeting of a future California CCHD chapter with some fifteen people, including a few HD family members, followed a month later by another meeting, this time with the neurologist Charles Markham to talk about Huntington's disease to a much larger audience.[23]

From the start, my father's vision differed from Marjorie Guthrie's in New York. Rather than a patient advocacy organization whose members typically had Huntington's in their families and were desperately in need of services, he imagined something more like a public foundation to fund and support science. With a few exceptions, donors to the California CCHD chapter, at least initially, did not have Huntington's in their families and were not "members" unless they served on the board of trustees or the science advisory board. While Marjorie wanted to lobby Congress to support research and expand social services, Dad wanted to interact with scientists and have a voice in choosing which research to fund; for example, he wanted to focus on basic scientific research, that is, research that might not be immediately relevant to treatment. He did not want the California chapter to fund clinical research or research on care. Despite his lack of scientific training, he grasped early on that the field of molecular genetics was changing all of biology, that knowledge about the biochemistry of the brain was advancing rapidly, and that offering scientists not only funding but also access to patients and tissue samples would be a critical means of getting them involved.

As a Freudian psychoanalyst with no background in biology, Dad might have hesitated even to try to understand the neurobiology of Huntington's. However, his experience with schizophrenia—and briefly with muscular dystrophy—probably made the subject less intimidating than it might otherwise have been. He was comfortable reaching out to neuroscientists such as Eugene Roberts, who shared an interest in schizophrenia, and then asking for help with Huntington's. Roberts, in turn, admired my father's psychological and social skills and was impressed by his strategies for getting good scientists interested, including complete novices who had no background with the disease.[24]

In those early days, Dad aimed to identify creative thinkers willing to join a scientific advisory board that could oversee the hiring of an investigator to head a major Huntington's research center, possibly at Cedars-Sinai Medical Center or UCLA. He sought out investigators who were experimenting with the basics, as he and Henry had agreed. And indeed, in March 1970, Linus Pauling, who had no expertise in Huntington's, agreed to join the scientific board, an acceptance that encouraged Dad to reach out to other eminent basic scientists such as William J. Dreyer at the California Institute of Technology, who joined in April 1970. Drawing on Menninger's multidisciplinary approach, Dad wanted the science board to represent a "broad spectrum," not only physicians with clinical expertise but also biochemists, molecular biologists, pharmacologists, and the like who may never have seen a person with Huntington's but understood new technologies. Already my father was pitching Huntington's as a model for understanding other neurological disorders and an ideal disease for the study of aging, an approach that would be central to his strategy going forward.[25]

In his journal Henry offered glimpses of his brother in late 1968 when Dad came to the East Coast for a CCHD meeting, "looking older, hair grayer, face lined, and less ebullient than usual." Nonetheless, his talk had "some of the qualities of the old-time revival meetings," Henry wrote, commenting that Mom had given Dad a cause and a crusade. "Promising that the psychological moment is just around the corner," Dad had proclaimed his intention to reach out to "every neurologist, chemist, molecular biologist, physicist, and even computer expert within range of our organization" to get them involved in the problem of Huntington's. He then told the audience his version of an old Jewish tale that he would repeat in different contexts. The story "might seem chauvinist," Dad warned, but although he was a Jew, he was also a Christian and a Muslim, black, red, and yellow. The story went as follows: "A rabbi, a priest, and an imam all received a visit from the Angel of Death warning them of a great flood to come. The Christian and the Muslim leaders called their flocks to the church and the mosque to pray. But the rabbi got the scientists together to try to solve the riddle of living underwater." According to Henry, Dad's appeal was toward "money raising, towards rousing the public to help in all kinds of ways, towards influencing politicians, towards not hiding the illness, letting children know."[26] "Learning to live underwater" became a kind of mantra, one that my father would draw upon when he faced difficult physical challenges later in his life. I now draw on it myself.

Soon after establishing the California CCHD chapter, Dad began the project of creating a board of trustees willing to fundraise, make financial contributions, bring visibility to the chapter, and establish the chapter's authority. Somewhere along the line, the trustees agreed—or did not object—to the idea that they would pay all administrative expenses so that the chapter could promise that 100 percent of monies raised would go toward research. Dad took this pledge seriously and managed to sustain it for many years, thanks to his and Maryline's willingness—and later that of Nancy and various volunteers—to shoulder many administrative tasks, thereby keeping expenses low. Moreover almost all trustees had significant financial, social, and cultural resources to contribute. Or as Dad later put it, they "represent a very influential segment of the community."[27] The board grew rapidly. From five members in early January 1970, it had expanded to nine by May 1970, to fifteen by April 1971, and to thirty-six by March 1972.[28]

And it was here that Dad ran into the ethical dilemma of a prominent clinician who raised money for his own foundation's biomedical research. Of the thirty-six trustees in 1972, at least eight were patients, former patients, or the spouses of such. Like Marjorie Guthrie in New York, Dad had many high-profile connections in Los Angeles, in the visual arts, in Hollywood, and in such associated professions as law, accounting, and public relations. The problem for Dad was that many of his connections were made through his practice; that is, they were his patients or the family members or friends of patients. He had previously socialized at openings with some of the artists in his therapy group, and he had met and socialized with art collectors at these openings. But the start of his advocacy for Huntington's research changed the metric dramatically. The celebrity and wealth of many of his patients and their friends were an obvious boon to publicity and fundraising. But did my father's patients and ex-patients step up willingly to contribute funds to his personal cause and to host parties and meetings at their homes? Or did they feel pressure to contribute? Did they seek approval or favoritism by donating money? And what about those who agreed to become trustees and, later, join the board of directors? Were members of "Group"—a high-power therapy group—motivated to contribute out of competition with fellow Group members, a desire to appear generous, a bid for favoritism, a wish to "belong"? How much pressure did they feel to participate in what Dad increasingly described as the "foundation family"? Some did feel uneasy, questioning his power to make patients feel they ought to contribute money or contribute in some other way, even if he did not do so intentionally. Some felt that Dad tended to play favorites, preferring those who gave large donations. Other analysts also questioned his blurring of the boundaries between the professional and the personal.[29]

Of course many celebrities donate their talents to causes they support. Many clinicians from the 1980s on, in the United States at least, approached their wealthy patients for contributions to medical projects, whether a new hospital wing, a new clinic, or research. Were psychoanalysts and psychologists any different? In interviews later in his life, my father sometimes claimed, disingenuously in my view, that the participation of certain of his patients as foundation board members or as financial supporters helped *them* because they were making possible great science. He was giving *them* an important cause, enabling them to participate in state-of-the-art biomedical research. But such self-serving claims did little to address the ethical issues involved. Dad had once been scrupulous about not blurring the boundaries between his social and professional lives. While Nancy and I were growing up, he would avoid parties where a patient might be present. He had socialized with some of his artist patients, but he was not involved in advocacy at the beginning; he was not raising funds for his own cause then. It was as if the intrusion of Huntington's into his life made all such ethical rules irrelevant. He had a larger mission to accomplish. As his daughter, I found the situation uncomfortable because I knew Dad was motivated by his love for Nancy and me. I benefited from his violation of boundaries. They were not my boundaries; they were not my patients. But it troubled me nonetheless.

One tragic event does not preclude others. In February 1969, while Dad was struggling to come to terms with Huntington's in our family and to organize the California CCHD chapter, his beloved patient John Altoon died suddenly at a party from a massive heart attack. His death was a profound loss not only for the Los Angeles art community but for my father, who had loved Altoon like a son. Dad was not given to lyricism, but a few days after the funeral he sent Nancy and me an emotional two-page, single-spaced letter describing what had happened. "Just before he went to the party," Dad wrote, "John phoned me to say that he had been laying some pretty heavy things on me recently and that I deserved to know when he was feeling well. In fact he felt great. He told me he wanted me to come over sometime next week to pick out two drawings to send to each of you. He thought it would make you feel good and happy too."[30] Dad explained that they had arranged to meet the following Tuesday but that Dad had made an appointment that day at Cedars-Sinai, so John was going to drive with him to keep him company. "He enjoyed the idea because we would have a longer time together," Dad added. "Then I said he should have a great time at the party, and we hung up." Later that evening Dad got a call to come to the home of Betty Asher, a local patron of the arts, where John had just died. "I don't have to elaborate on my feelings or the things I had to deal with in the next few days," he said. He estimated that the funeral was attended by at least 350 people and possibly many more. "John's widow, Babs, plus all John's friends were absolutely against having a regular service with any minister," he wrote. "So I suggested that we just let people stand up and say whatever they wanted for the full hour of the ordinary service. They loved that idea and I agreed to lead off with a statement that would be my own personal eulogy. Then I invited people to stand and speak as the spirit moved them. My own eulogy was, I am told, so beautiful, that many just broke down and cried."

> And when I was finished I invited them to stand and speak. And one by one they did for the full hour. And the types and faces were extraordinary. Blacks, white, yellow, hippies, the rich, the poor, the most unusual mixture of people that has been assembled anywhere. And some told simple stories about John, and some praised him, and some wept for him, and some told funny, laughing stories about him, and the people cried and laughed and it was just beautiful, the way John would have loved it. And then at the end I said that this was like John's religion, filled with poetry and tears, and laughter. And I told them that many knew him as a painter but few knew him as a poet and that they could all come to the graveside and there I would read just a few of John's poems which I had collected from his scrapbooks over the weekend. And that's what we did. [We] went to the grave and there I read three short poems he had written which were so apt they were tremendous.[31]

Dad remained in touch with Babs for the rest of his life, determined to help promote and sell John's work, which they both felt had not received its due recognition.[32] He wanted Nancy and me to appreciate that John "was such a unique man that his very life is a memorial and no one who ever met him will forget him. One of a kind and that is so certain that you must treasure every memory you have of him." A few years earlier he had summed up what he most admired in this memorable patient and friend: "One can say with full confidence that Altoon has never shirked, either in his work or his life, an open and honest receptivity to the deepest pain and the wildest beauty that life has to offer."[33]

I I

Workshops of the Possible

*I found in science a mode of playfulness and imagination, of obsessions
and fixed ideas.*

—François Jacob, *The Statue Within*

I n his autobiography, *The Statue Within*, the biologist and Nobel laureate François Jacob contrasts two types of science. Day science "employs
reasoning that meshes like gears, and achieves results with the force
of certainty." It advances in light and glory. Night science "wanders blindly. It
hesitates, stumbles, falls back, sweats, wakes with a start." Night science feels its
way forward, questions itself, it is a sort of "workshop of the possible."[1] I doubt
my father ever read *The Statue Within*, but he would have loved this passage if
he had. For what he aimed for in the workshops he soon developed to galvanize
research on Huntington's was akin to night science. Night science resembles a
kind of psychoanalysis, a mode of scientific free association.

At first, however, his mind was focused on daylight concerns: raising money
for a Huntington's research center and establishing the California CCHD
chapter on a firm financial footing. A Woody Guthrie tribute had taken place
at Carnegie Hall in New York shortly after Guthrie's death in October 1967.
Now the legendary producer of that event, Harold Leventhal, offered to produce "without fee or charge" a second tribute in Los Angeles at the Hollywood
Bowl, with proceeds to go to the California CCHD chapter and Huntington's
research. With his record of representing such progressive artists as Pete Seeger

and the Weavers, Leventhal had access to many great performers, as well as to the writer Millard Lampell, who had written the script for the Carnegie Hall tribute and was willing to do this one, too, gratis. Soon Leventhal had lined up Joan Baez, Judy Collins, Ramblin' Jack Elliott, Will Geer, Arlo Guthrie, Richie Havens, Country Joe McDonald, Odetta, Earl Robinson, Pete Seeger, and others, all of whom agreed to waive their fees. Dad was thrilled, as a successful concert at the Bowl, with its eighteen thousand seats, could help launch the research center he dreamed of.[2] He was convinced that a record and movie of such an event could bring even larger returns than the concert. Though aware that Guthrie's autobiography, *Bound for Glory*, was being adapted for the screen and that a feature film could conflict with a documentary of the concert, he was willing to take the risk. Several trustees of the California CCHD chapter donated funds to film the concert, and Warner Brothers agreed to put an additional $20,000 toward filming and recording.[3]

Dad took on an enormous amount of work for the concert, at times expressing an almost manic exuberance that not only Huntington's but all genetic problems could be overcome. We were in the age of biology, he wrote to me. He wanted to give it a big push.[4] But as the event grew closer he became anxious, knowing that there were "so many ways to get wiped out in the theatre." Dad was living by himself in Frank Gehry's recently built Hillcrest Apartments at 2807 Highland Avenue in Santa Monica, a modest building with surprising angles and inventive details hinting at Gehry's future directions. Frank had joined the board of the new California CCHD chapter and was helping to secure permits for the concert. He also lived in the building, in the apartment above Dad's, and he recalled hearing my father typing away at night that summer, windows wide open, sometimes as late as midnight. It was not an ideal arrangement for a therapist and his patient, and Frank was not entirely comfortable with it. Fortunately, despite Dad's pleasure in the apartment, he decided he needed more space and after a year moved to a larger if more conventional place.[5]

The concert, on September 12, 1970, exceeded all expectations. Even rehearsals on the top floor of what was then the Continental Hyatt House on the Sunset Strip—which Nancy and I managed to sneak into—were memorable. Here were the great Odetta belting out "Ramblin' Round Your City"; Joan Baez sprawled on the floor, as if paying homage; Country Joe with his saucy demeanor; tall and lanky Arlo Guthrie; Richie Havens and other performers, musicians, technicians wandering about; Marjorie holding a grandchild, little kids underfoot.

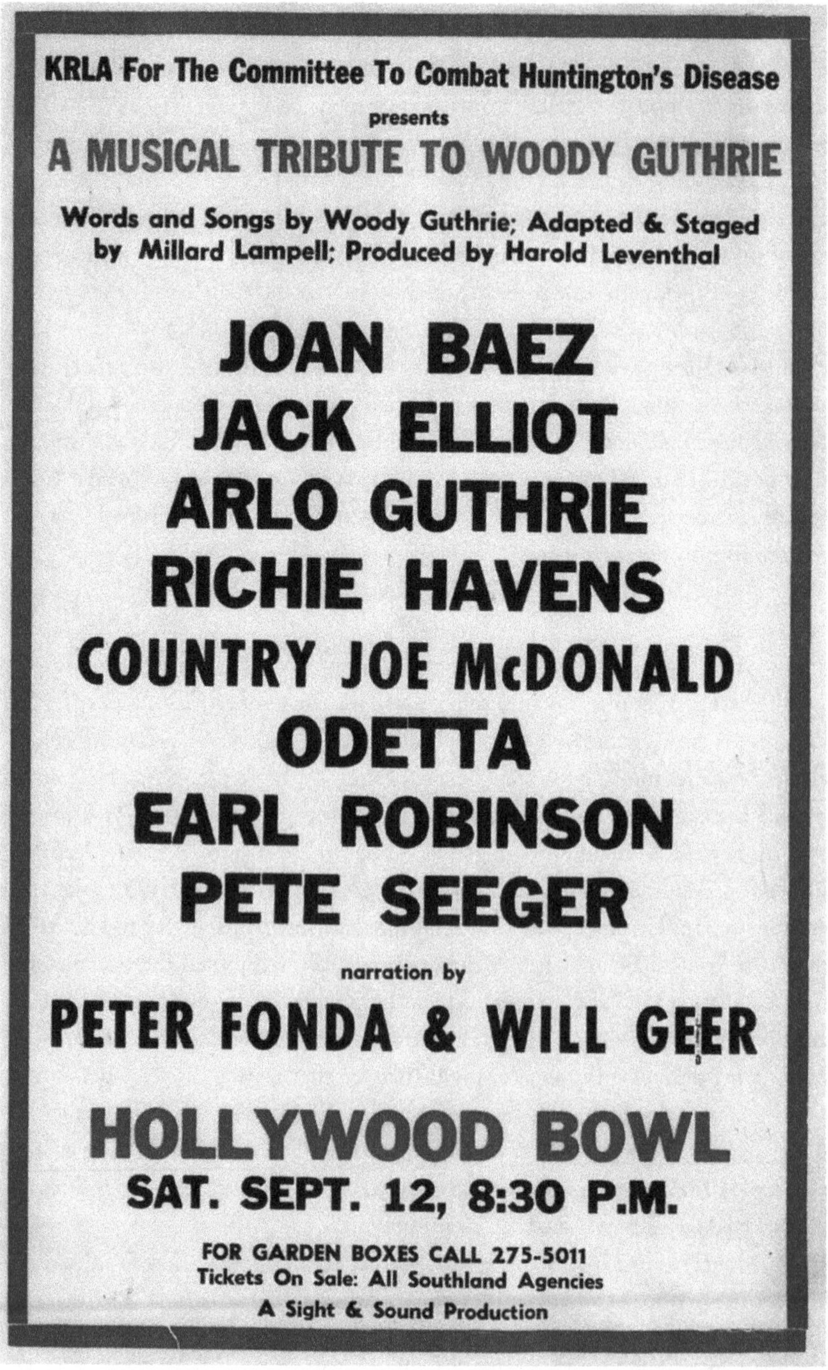

FIGURE 11.1 A poster for the September 1970 Hollywood Bowl Tribute to Woody Guthrie.

Courtesy Hereditary Disease Foundation.

Dad was jubilant, even before all the proceeds were in. When they eventually totaled $133,000, he felt certain that the California CCHD chapter was on its way to establishing his dreamed-of center. He wrote emotional letters of appreciation to those who had worked on the project, especially to Marjorie, who thanked him effusively "for succeeding at what seemed like an impossible project." "It isn't only the financial success," she added. "The whole spirit of your leadership and relationship with your community is, undoubtedly, a great 'inspiration' to all of us."[6]

As happened so often in my father's life, his dreams soon confronted frustrating realities. It was part of his genius, I believe, that he was able to take in deep disappointments and setbacks without getting derailed from his larger objectives. He was disappointed to learn, early in 1971, that a feature film, coproduced by Harold Leventhal through United Artists, was now in production and was indeed diminishing interest in a concert documentary. (*Bound for Glory* would be released in 1976, garnering several Academy Awards). Dad was furious, however, when he discovered that after he and Levanthal had signed an agreement with Warner Brothers that the Bowl film footage and concert recordings were the property of the CCHD California chapter, Warner Brothers had signed a separate agreement with Leventhal and CBS to produce a record album combining music from the Hollywood Bowl and Carnegie Hall tribute concerts. Proceeds from this album would go, not to CCHD and Huntington's research—which was the sole reason for Dad's efforts—but to a new Woody Guthrie Tribute Fund (later called the Woody Guthrie Foundation) whose primary purpose was to create a museum and support scholarship on Guthrie and American folklore and music, with funding for scientific and medical research a lower priority. While the California CCHD chapter would keep all the income from the concert itself, Dad saw his hopes for more returns from a film and record evaporate overnight. Perhaps the possibilities were always uncertain, for the California CCHD chapter's rights to use the songs for film and television depended upon approvals, releases, and clearances that had not yet been secured. In the end, the concert footage disappeared into the vaults of Warner Brothers, where it remained for nearly fifty years, until after Dad died, when Woody's daughter, Nora, rediscovered it and the cinematographer Fred Underhill, who had overseen filming of the concert, turned it into the documentary film Dad had always dreamed of, the "Woody Guthrie All Star Tribute Concert 1970."[7]

And then new disagreements arose with the national CCHD office in New York having to do with organizational structure and whether CCHD would become a decentralized association with autonomous regional chapters, as Dad wanted, or an organization with centralized control, as Marjorie proposed.

Eventually, after many meetings, much correspondence, and a number of conversations about the issue, my father concluded that it was better for the California CCHD chapter, with its fundraising firepower, "to separate in peace and harmony" from the national organization and for each to pursue their shared aims from their different vantage points.[8] Staying together at all costs— in an organization, as in a marriage—was not in Dad's nature. However, he was determined not to let his anger get in the way of the larger goals he shared with Marjorie, of whom he was genuinely fond. They would work together as equal partners, not as one subordinate to the other. I think Dad worried that because many of the California chapter supporters were motivated primarily by personal loyalty to him, any diminution of his authority would weaken their commitment. Perhaps Marjorie in New York felt the same.[9] The California CCHD chapter eventually morphed into a new independent entity, the Foundation for Research in Hereditary Disease, to be renamed a few years later the Hereditary Disease Foundation.[10] It was the end of a long struggle, but it was not the end of a relationship. For within a year, Marjorie and Dad would serve together as honorary co-chairs of the landmark Congressional Commission for the Control of Huntington's Disease and Its Consequences, which had a mandate to survey the needs of families with Huntington's across the country and to make policy recommendations based on their findings. And Nancy, as the commission's executive director, would be their boss.

Even as Dad worked on the Hollywood Bowl event, he continued to solicit advice on the research center. He soon discovered that the established Huntington's investigators opposed this idea, arguing that there were already many promising research ideas in need of grant funding. Nor were younger scientists enthusiastic. They too favored funding grants and also postdoctoral fellowships as a means of introducing new investigators to the problem of Huntington's. Within a few months my father became convinced that whatever advantages such a center might have from a financial perspective would not make up for the disadvantages from a scientific point of view. Dad had a vision, but he was also a pragmatist. Eventually the California CCHD chapter's two boards, the science advisory board and the board of trustees, agreed that a center was not economically feasible or scientifically advisable at that time.[11]

One idea that everyone did support was a small, freewheeling, interdisciplinary workshop, a different version perhaps of the groups my father had been conducting with artists for a decade. On a trip to Washington, DC, and New York in December 1970, Dad had found both the Huntington's experts and CCHD's national board enthusiastic about such a workshop as a preliminary to the big March 1972 Centennial Symposium being planned for the one hundredth anniversary of George Huntington's landmark paper. He immediately began soliciting names of potential participants from John Whittier and Ntinos Myrianthopoulos. Besides inviting some of the senior clinicians involved in the World Federation of Neurology Research Group on Huntington's Chorea including Charles Markham and André Barbeau, he sought out basic scientists such as Nobel laureate Julius Axelrod of the National Institute of Mental Health, the eminent neuroscientists Seymour Benzer and William J. Dreyer from Caltech, and Isabel Tellez-Nagel from the Albert Einstein College of Medicine, all of whom accepted. Scientific meetings that included investigators from different fields of biology were not common at the time. Nor were meetings that mixed clinicians with basic science researchers or experts with novices, in this case those who knew nothing about Huntington's but were good scientists and would look at the problem "as it should be looked at."[12]

The first workshop, in early March 1971 at the Santa Ynez Inn in Pacific Palisades, a few blocks from the Pacific, met with resounding success. Afterward, Dad was ecstatic, his letters bursting with enthusiasm for the great young people in the field, new developments in genetics, and the "consensus among all those present that this was one of the best organized and most stimulating scientific meetings any of them had attended in recent years."[13] So much so that André Barbeau thought that many of its elements should be incorporated into the upcoming Centennial Symposium; also that some of the same attendees should be invited to participate, including Dad himself, whom Barbeau asked to chair a panel. Dad was thrilled at the lively interdisciplinary exchanges at the workshop, both about scientific issues and about organizational strategies for advancing HD research. He was heartened by responses such as that of the neuroscientist Elie Shneour from the City of Hope National Medical Center, who wrote to him, "I have never seen clinicians interact with such effectiveness with so-called basic scientists and meet on common ground." Shneour had come "almost completely ignorant of this baffling disease, and I left no less ignorant, but permeated with curiosity and interest. No matter what work I may do in the future, it cannot [help] but be influenced by what I learned at that [workshop]."[14] This was precisely what Dad wanted to hear, and it strengthened his commitment to creating an organization focused on science with an emphasis on seeking out, supporting, and funding young investigators—including women like those at the first workshop—at the start of their careers. He was more than ever convinced that laypeople—nonscientists—could directly influence scientists and that encouraging scientists' interests and enthusiasms and offering funding and resources for their projects, for example access to patients and biological samples, could have an impact on the future of the disease.[15]

Still, he was keenly aware of how much he did not know and deeply grateful for the interest of those who attended the workshop. He wrote effusive letters to some of the attendees, thanking them for being "so damned kind, listening to all our plans, proposals, etc., and responding as objectively and as reasonably as you could."[16] He even seemed to enjoy the challenge to his psychological expertise and diplomatic skills in dealing with all the different perspectives and interests he encountered, although "I'll have to be a Disraeli to maneuver through those waters," he told us. "Already pressure on me from all sides."[17]

For my father, the next two decades were years of dizzying activity as the California CCHD chapter and its successor, the Hereditary Disease Foundation, expanded its program to include as many as six workshops a year, a dozen grants and postdoctoral fellowships, and all the ancillary activities, particularly fundraising, to support these initiatives. For a while he found himself in the position of acting science director. He was often the one who telephoned and wrote letters to scientists, solicited evaluations of potential grantees and science board members, acted as an intermediary between investigators, reported on available state and federal grants and on relevant impending legislation, developed plans for future workshops, hired nurses to collect biological samples, announced the results of experiments, and kept discussion open and amicable even when it went into emotional territory. He helped develop the foundation's scientific priorities and practices, partly through an extensive process of seeking advice. Dad put a lot of effort into figuring out whom to listen to and whom to invite onto the science board and to workshops. While initially he was happy to find anyone who was interested in Huntington's or who could become interested, over time he became more selective. He could be tough in judging those whom he felt were not creative or communicative enough, those who were smart but too quiet or too talkative or too dour and humorless. He reacted negatively to people whose voices grated or whom he could not hear or who interrupted and disrupted the flow of ideas. He wanted people with imagination and boldness, who knew how to listen, who were not defensive, and who did not attack others. He always wanted to know who was "the best" in any field, an attitude that sometimes annoyed me in its frank elitism. However, "the best" did not necessarily mean having Ivy League or other high-power credentials. It was an ineffable quality that he thought he could recognize but not always define. The right mix of perspectives and personalities was crucial in determining the success of a workshop, and it took all my father's psychological and social skills to come up with an optimal blend.

At the early workshops Dad was an influential presence, a "benign father figure," in the words of Eugene Roberts. These gatherings of about fifteen to twenty participants, held typically in hotel meeting rooms or at universities around the country, soon became incubators of new ideas, generating research proposals and requests for funding from attendees. For a long time those held in Santa Monica included a celebrity-studded Saturday night party or dinner, usually at a restaurant but sometimes at the home of board members such as Jennifer Jones Simon and Norton Simon or Julie Andrews and Blake Edwards, where young scientists from Podunk, as Eugene Roberts put it, mingled with artists, movie stars and millionaires.

Dad liked shaking up the standard format of scientific meetings by banning slides, formal presentations, and reams of data. He would tell workshop participants that he believed in free association and wanted them to speculate and let their imaginations roam. "All the academics—they didn't like it, until they got into it," he told me later, "and then they loved it."[18] He would tell participants that the purpose of the workshops was not the exchange of information so much as generating hypotheses for research. He was thrilled to see "young people bringing new technologies to the older people and the older people bringing their wisdom and experience to the craziness of the young people." He paid attention to the emotional atmosphere. "Very early on I had the sense that the element of play was needed," he said. He wanted people to have fun and believed that good ideas emerged that way. Dad did not talk much at these meetings. When he did, he spoke low and quietly. He wanted to create an atmosphere of being "at home."[19]

The growing reputation of the Huntington's disease workshops spun off other invitations, possibilities, and projects for my father, including one at the Jet Propulsion Laboratory, affiliated with Caltech in Pasadena, on the potential for the laboratory to establish an institute of genetic research. Leroy Hood, a Caltech geneticist, and his group had developed the prototype for a machine that could sequence genes, thereby enormously speeding up a process that had been slow and laborious. Dad was keen to have Huntington's disease on this group's research agenda and accepted immediately when he and Nancy were invited to see the "gene machine."[20]

A few years later, the Frederick R. Weisman Art Foundation, no doubt at the behest of Frank Gehry, invited Dad to moderate a workshop on the relationship between art and architecture, which was attended by sixteen of the most eminent architects, artists, and art critics in the world, including Frank. Dad introduced the workshop by explaining that his past experience had led him to believe that real creativity is often the result "of casual conversation and free association among people in the same or related disciplines." He said that he hoped that the discussion would remain at the level of casual conversation.[21] Through Frank, Dad was also once invited to the famous annual International Design Conference in Aspen as a "critic at large," where he unexpectedly found himself speaking about genetics to an audience that included the great British geneticist Conrad Waddington, credited with developing the concept of epigenetics (the field of study focused on the factors that influence the expression of genes, including toxins, hormones, enzymes, and much else). Waddington happened to be a last-minute invitee, Dad reported to Nancy and me, "so I introduced him, explaining it was preposterous that I should lecture on these subjects with him in the audience." Dad told the audience he'd give a journalistic view, "which I did[,] and Waddington was most complimentary."[22]

All this activity seemed to energize Dad, even if at times he felt exhausted. "Wow! Am I inundated," he wrote to Nancy and me in a typically breathless missive. "With patients. With the HD foundation. With prostituting myself all over the lot."[23] His letters to us were filled with exclamations of how "super busy" he was—his "wingding schedule," "pressures tremendous," "things as usual here— hectic," "many meetings coming up and I'll be snowed for a month or two"—and with apologies for joint letters, which nonetheless were sometimes single-spaced, typed dispatches of one to two pages. He was so busy that he was walking around with holes in his socks. "And I used to complain about the pressure of practice!" he once wrote. "Seems like a Utopia lost in the face of Foundations and Commissions and politics and fundraising."[24] But even his most rushed letters managed to include an upbeat message. "We begin to get much coverage, more research and interest," he wrote just a year after Mom's diagnosis. "Don't need vacation. Need to clean things up."[25] He was beginning "to see some areas in which there is real hope in HD." He wanted to get a think tank going to bring molecular biologists, enzymologists, and other specialists together to brainstorm.[26] He thought many of the doctors in the field were handicapped in some ways, "and I just begin to perceive where they really need help. And I propose to try to get it to them." He was excited by recent discoveries of missing enzymes in six hereditary diseases, some of which were "close to HD. What a breakthrough! I want these fellows on our team! I've written them," he told us. "The next step will be to figure out how to feed the enzymes in. They have such a project in formation. If it works—then WOW!"[27]

When his brother expressed concern that he was driving himself too hard, Dad agreed but explained that his frenetic pace had psychological benefit. "You are right that I work under extreme pressure," he told Henry. "And I feel it now more than ever. But I've never done otherwise so long as I can remember. Just that it makes less sense in view of my age. But I'm inclined to keep on going and burn it all up. Might go out in a flare instead of slowly and painfully," he admitted. "And I do have some real fun and kicks this way."[28] Like going to art openings, including that of the light-and-space artist Larry Bell, who "is not a patient but seems very successful anyway."[29] Or cohosting a fundraiser for the Black Panther Eldridge Cleaver's defense fund with a former patient. Or riding a bike along Santa Monica beach on New Year's Day with Nancy, Frank Gehry, Ed Moses and his wife Avilda, and the sculptor De Wain Valentine: "Just about the best way to get the old year over I know."[30]

Nancy, too, worried about Dad exhausting himself and contemplated a return to Los Angeles to take over running the foundation. But Dad reassured her that he wouldn't enjoy himself if he didn't "tangle with this problem." And for a long time he felt certain that he could help get it solved. "What the hell else do you live for than getting your teeth into something meaningful and shaking out a drop of real nourishing juice?" he asked. He didn't think he'd be a different father if he were less involved with Huntington's. "I'm by nature involved," he wrote to my sister. "If it weren't this I'd probably be tackling all those research invitations on schizophrenia that the NIMH dangles before me. Or work my ass off with patients. I'll live longer this way than taking it easy." He did want more time to write. But he was going to do that anyway. Someday.[31]

While he had ceased taking month-long summer holidays in Europe, he used international Huntington's meetings in such places as Brussels, Copenhagen, and Leiden as a base for a few days or a week of vacation. He stayed at a chalet in Gstaad, Switzerland, holed up to write at a house on Fishers Island off the coast of Connecticut, and took his writing pad up the Pacific coast and to Honolulu in Hawaii. The entrepreneur and philanthropist Max Palevsky invited him for a week aboard his yacht sailing in the Caribbean along with the writer John Hersey and his wife, Barbara, and the playwright Lillian Hellman, who was becoming a friend.

And he continued to correspond with HD family members who wrote asking for advice. Though he emphasized the horrors of Huntington's in public statements and fundraising appeals, his personal letters to family members offered a more hopeful message and encouragement for the future. He had never forgotten Henry Sternberg's daughter Linda, now deceased, whose resourcefulness inspired his deep admiration. He kept in touch with her younger sister, Ilse,

twenty-two at the time, also born with muscular dystrophy and also living a "marvelously full and exciting life" while using a wheelchair and being vulnerable to respiratory challenges. Dad admired her family's straightforward approach to the illness and their philosophy of living with disability. As he wrote in a 1971 talk for the Muscular Dystrophy Association, speaking of Ilse, "It was as if everyone said, 'We are all afflicted in one way or another. We are all going to die. We all have our own limitations. We must all live within those boundaries. But we must live our lives, not die within them.' And this attitude seems to have sunk deep within her soul. So she is matter of fact, and therefore comforting, even when she deals with me and HD." The Sternbergs and their daughters taught Dad not only about caring for a child with a disability and about living with disability but about living, period.[32]

He carried that point of view into his correspondence with Huntington's families. "From all that I have heard, and I have heard a great deal," he wrote to one desperate father of a child recently diagnosed with Huntington's, "the course of the illness does not necessarily or inevitably follow such a dramatic and tragic course." He urged the father not to buy the "horrifying, dire, tragic picture" that had been painted by a genetic counselor and advised that medications could make symptoms somewhat more manageable. Always he tried to convey hope and comfort in the fact that "top-level" scientists were working to find answers. "Someday I do believe we will celebrate the answer to this desperate problem," he concluded one letter. He believed in the possibility of finding treatments for schizophrenia, and he carried that hopeful attitude into his talks on muscular dystrophy. Now he was applying it to Huntington's as well.[33]

Mom was dying. It had been ten years since the morning in 1968 when a police officer stopped her as she crossed an intersection in downtown Los Angeles, on her way to jury duty, and accused her of being drunk. Ten years during which Mom suffered as much from loneliness as from her steadily declining abilities. Dad had remained in charge of her care throughout. Nancy and I visited, wrote letters, and made phone calls, and we tried to encourage Mom and keep up her spirits, but we lived far away. We did not have to make the tough day-to-day decisions that inevitably came up over the course of her lengthy illness. We did not have financial responsibility for her. When we visited her at the nursing homes where she spent her last years, we had time to take her out to eat while she could still manage restaurants, take her for walks, listen to music or watch TV with her, or just sit with her and keep her company. Toward the end, she was in a Santa Monica convalescent hospital, as it was called, strapped to her bed to keep her emaciated arms and legs, which were in constant motion, from injuring her. The place was decent enough, and the overworked staff tried to be gentle and kind. Dad kept in touch with Bernie Salick, who was acting as Mom's primary care physician. She died alone on Mother's Day, May 14, 1978. She was sixty-three years old.

With Mom's death, Nancy and I—and Dad, too—felt free to remember happier parts of her life, the smart and lively young woman she once had been, and her unstinting love for us, no matter how weighed down she felt with sorrow, loneliness, and fear. Nancy and I composed our own grief rituals, each afternoon driving up the Pacific Coast Highway from Dad's apartment to walk on the beach or hike a new trail in the Santa Monica mountains, where we talked about Mom's love of nature and tried to recall the names of the flowers and shrubs she once had known. We cried together, not for her death but for her life, for the mismatch of her marriage and for the illness that had shadowed her from her youth and taken first her father, then each of her brothers, and now her. We cried for her terrible loneliness, for the stigma and silence that surrounded Huntington's, and for the great promise of her college and university years that had never had the opportunity to flourish.

That summer my partner, John Ganim, and I traveled to Hawaii to visit friends. We had met shortly after Mom's diagnosis while we were both in graduate school at Indiana University. His roommate and I had been teaching assistants together, and John and I met one afternoon when I stopped off at the house they shared. John immediately won my heart when he made me an espresso with his new espresso machine and told me about the film society he had organized as an undergraduate at Rutgers University and about the underground films they had showed. He was warm and smart and good-looking, with a great sense of humor and an original way of looking at the world. I was captivated. Soon I fell in love. And then a few months after we met, John was drafted into the army, which thankfully assigned him to a base in nearby Fort Knox, Kentucky, instead of Vietnam. With a master's degree in English, he ran a language laboratory and edited documents while writing occasional articles for an underground antiwar newspaper on the base started by fellow recruits. Somehow we made it through two difficult years, and once he was released, we began living together and finishing our PhDs, his in medieval English literature and mine in Latin American history. We each got tenure-track academic jobs, mine at Sonoma State College (now University), an hour north of San Francisco and John's at the University of California, Riverside, some sixty miles east of Los Angeles. Commuting between Riverside and San Francisco, I had tried to see Mom as often as I could. And now she was gone. I wondered who would be next.

In August that year, Dad turned seventy, an age when many men are thinking of retirement. He was thinking of writing, of friends, and of genes. And then Henry died of lung cancer in December at the age of seventy-two. Although they had not been as close recently, my father had never forgotten how Henry had been his lifeline and support at the most difficult moments of his life. If Dad was often critical of his older brother, disputing his optimism on some topics (for example, that Nancy and I might both escape HD) and his pessimism on others, for a long time he depended on Henry as his most trusted and intimate confidante. Henry had borne most of the responsibility for their mother during her last difficult years with dementia. And now he, too, was gone.

As Dad often did when hit by a major loss, he took a step forward. The scientific director of the foundation was leaving for a new job on the East Coast. So in late 1978 my father hired a talented young biophysicist in his thirties, Allan Tobin, to take his place. As a postdoc at Harvard, Allan had participated in some of the early foundation workshops and impressed my father with his acuity and intelligence. He had recently accepted a position as an assistant professor in the biology department at UCLA, and he was willing to act as the foundation's scientific director on a part-time basis. Although Allan had no background in Huntington's research, he had imagination and energy and familiarity with the new molecular genetic technologies and those who used them. Dad thought he could carry this knowledge into the foundation's workshops and recruit young investigators who could bring them to bear on HD.

My father also turned to friends, including his new friend Lillian Hellman, with whom he carried on a playful and flirtatious relationship until her death in 1984. They met early in 1978 when the novelist and playwright Peter Feibleman brought her to Dad's office for help coming to terms with her deteriorating vision. Feibleman, who grew up in New Orleans like Lillian, had known her since he was ten years old when his parents introduced him to the thirty-something Hellman. Much later, they became friends and then lovers, although the romance was more Lillian's than Peter's, and this inequality of desire, and Feibleman's relationships with men, would be a continuing complaint in Lillian's letters to Dad. She lived in New York and in Martha's Vineyard during the summer, but sometimes in winter she visited Los Angeles, where Feibleman was living and where he introduced her to Dad.

My father was entranced. He had admired her play *The Little Foxes* that had opened on Broadway in 1939. Meeting Hellman awakened his memories of long ago dreams of writing plays himself. He was happy to talk with her about the psychological challenges of coming to terms with her loss of vision—a loss he would face within a decade. But he wanted to see her not as a patient but as a friend. After she returned to New York, the letters began—interspersed with periodic visits. She hinted that he could come to live with her in her summer house on Martha's Vineyard and advised him that, in the wake of Mom's death, "a new place, a new house like the Vineyard might be good for you (and better for me)." "I miss you in a most selfish way," she confided. "It did me a great good to see you and now you are so far away. But I send love and thanks."[34]

Lillian seemed surprised a few months later, after Dad had seen her briefly in New York, when he apparently conveyed the idea that he felt guilty, as if he had somehow misled her about his feelings for her (or lack thereof). "I miss you and I felt bereft and I don't know why that would make you feel guilty," she wrote back. "Why isn't it just what it is, a tribute to the pleasure and interest and comfort of you?" She ended with "I send you love, no guilt for sending it, and the hope that you will come back."[35] Years later, Dad claimed that Lillian had had a tantrum when he told her that he was not in love with her and that it had never been his intention to make her think he was. According to him, she had misinterpreted their conversations as a sign of intimacy and a developing love affair. "It was Lillian's style to take over men, declare them hers, and even develop the deep conviction that they must of necessity share her passions," he declared in his memoir. In the case of Dashiell Hammett, he believed, this approach had worked.[36]

Without Dad's side of the correspondence, it is difficult to know just what he conveyed to her. Indeed, there is a hiatus in her letters to him from June 1978 to March 1979, but it is unclear if letters from this period are simply missing from his archive or if the two were estranged during these months. Had Lillian really believed that he was in love with her and become angry when he confessed that he was not, as he claimed? Lillian flirted with many men, in person and on paper. Her letters to Dad were probably not unique. Nothing in her available archive or in the biographies of Hellman indicate that she felt more than deep affection for Dad and gratitude for the solace she received from him and for his ability to refer her to doctors and therapists. As Feibleman put it, Dad was "the man she'd come to depend on when she needed a certain kind of advice."[37] She trusted and looked up to him. In any case, whatever contretemps transpired was eventually forgiven or forgotten. There are no references in any of her extant correspondence to an argument over disappointed love. She continued to thank him effusively in every letter, telling him "in my gloomiest minutes you are the bright light on what often looks to me a very dark road" and always sending "much love."[38] My father's fear

FIGURE II.2   Lillian Hellman and Milton at Martha's Vineyard, circa 1981.

of being taken over by women, his sense that intimacy threatened entrapment, was a theme that surfaced several times in his writing and in conversations—and one that I believe he never fully resolved.[39]

In the spring of 1980 Lillian invited Dad to visit her in Martha's Vineyard that summer, and he asked if Nancy, John, and I could come along, too. He knew that we would be thrilled to accompany him, but I think he may also have wanted us there to forestall any embarrassing misunderstandings. As it was, the visit was a success. Lillian's wit shone as fiercely as ever, and she kept up a constant repartee that turned even mundane tasks such as making dinner into a sort of theatrical performance, albeit punctuated by her occasional cringe-inducing references to "Chinks," "faggots," and "kikes," some of whom were her friends and employees. Our company apparently pleased Lillian, for she later wrote to Dad that "the nicest part of the summer was the Wexlers' visit."[40]

Afterward Lillian began writing to me, offering to help me publish an article on the American anarchist Emma Goldman, whose biography I was working on. She did help, though not without Hellmanesque miscommunications and complications along the way. I took for granted that she was courting me as part of her playful courtship of Dad, as she invited me into her plans to capture him.

"I am sure you already guessed that I love your father very much," she wrote to me in September 1980. "I don't think you and Nancy should make him marry me, because I really don't want to marry anybody in the world, I guess, even your father. But I think you should make him live with me the rest of my life. Do see what you can do, even if you have to chain him and beat him up. I might make a mean stepmother," she continued, "but I will try not to, and I will buy lobsters any time I can sell anything." Wishing me good luck with my book, she invited John and me to come again the following summer and stay longer, "even if you have to leave your father in chains."[41] She also encouraged Dad to write about his experiences and discoveries, "and that's the greatest compliment I can possibly pay you," she added, "since I don't think many people, at bottom, have much to say about anything."[42]

My father continued to enjoy Lillian's company and conversation, but he did not feel completely at ease in her celebrity milieu. "I keep wondering what I'm doing in this world," Dad confided to me one evening early in 1981 when Max Palevsky invited Dad, John, and me to spend a weekend at his Palm Springs house with Lillian and Peter. We were having dinner at Le Vallauris, a fancy local restaurant where Frank Sinatra was holding court at a table next to ours. "I enjoy it [this world] for a time," Dad whispered to me, "but then I want to pull out, to leave. It lacks intimacy. Lots of verbal fireworks but no real intimacy."[43] But he continued his lighthearted correspondence with Lillian. In answer to one of her joking marriage proposals, he wrote that she was much too young for him to marry, "but I'll be delighted to adopt you. If you wish for references on me as a father please advise." By this time Dad was having some vision problems of his own, with "one eye going sour and may need surgery" although not yet to the extent that she was experiencing. He was "sticking close to home and making no plans for travel." He advised Lillian that "one of us must be strong enough to lead the other around."[44]

Dad did not weigh in on Lillian's writing about the McCarthy period, Russia, Stalinism, the House Un-American Activities Committee, or other flash points of political controversy in her career, though he strongly opposed her lawsuit against Mary McCarthy and tried to dissuade her from carrying it through. But he did write about Lillian's relationship with the truth or what is called truth. If he measured Lillian by the narrow definition in *Webster's Unabridged Dictionary*, he would have had to call her a liar, he wrote long after her death. But it never occurred to him to think of her as a liar. His recollection of reading the manuscript that eventually became Lillian's published book *Maybe* suggests his more nuanced perspective. "I came away from reading the manuscript with a very clear notion that Lillian was determined to novelize her life," he wrote, "and that in some strange, subjective way she was manufacturing her life experience, and even her identity, in the same fashion that a novelist would go about creating a character for his story." He did not believe it was delusional or hallucinated; "it was merely imagined, and the imagination was given sanction and validity." He thought that "we are all busy manufacturing ourselves or our histories, and this world would be a very boring or a very hostile place if all of us suddenly became total truth tellers."[45]

And if this imagined history involved other people? Dad did not seem bothered by it. He found "a certain charm in the effrontery and the daring." Lillian had written a script and believed it was real. "I think that's how she lived much of her life," he wrote, "barely able to distinguish between the facts of her life and the imagination of her life. It was not psychosis. It was not lying. It was more like that dream life Freud wrote about." Yet when she imagined that he was in love with her—at least so *he* imagined—he became angry at her for "this totally groundless interpretation." Apparently Lillian's habit of shaping the world according to her imagination was interesting to my father as long as it did not involve him.[46]

I 2

Making Friends, Making Love

All stories are haunted by the ghosts of the stories they might have been.

—Salman Rushdie, *Shame*

My father and Nancy stand together, flanked on one side by the young Harvard molecular biologist James Gusella and by the Indiana University geneticist Michael Conneally on the other. Dad and Jim hold glasses of wine over a birthday cake streaked with ribbons of frosting in the shape of a double helix. They look reflective, as if anticipating the complexities to come. Nancy and Mike beam into the camera, euphoric at this rare moment of triumph in science. We were celebrating the discovery, in August 1983, of what scientists called a genetic marker for the Huntington's gene, in this case a short variable segment of DNA located at the top of chromosome four near the gene. Geneticists had previously used differences within families to track genes; for example, differences in blood type that correlated with a specific condition or disease. But using slight variations in DNA—known as DNA polymorphisms—to map genes was new. What Nancy, Jim, Mike, and the MIT geneticist David Housman, the principal architect of the HD gene mapping project, discovered was that, within the families they had studied, those individuals who had Huntington's usually had one pattern of DNA within this variable stretch of chromosome four, and those who did not had a different pattern. What made their discovery so valuable was that the variable segment was so close to the Huntington's gene on the chromosome that it usually traveled together

FIGURE 12.1 Celebrating the genetic marker discovery in 1983. Left to right: P. Michael Conneally, Nancy, Milton, and James Gusella.
———
Courtesy Hereditary Disease Foundation.

with the gene; the two were typically inherited together. In the geneticists' par-
lance, this segment was "tightly linked" to the gene. Three years after starting
their search, supported partly by the Hereditary Disease Foundation, they had
narrowed the territory of the Huntington's gene from the entire human genome
to a small section of chromosome four.[1]

Mapping genes requires large multigenerational families whose members are
willing to share their histories and donate blood and sometimes skin samples.
And it was Nancy who, in 1981, began leading a team of geneticists, neurologists, nurses, social workers, and students on annual month-long expeditions
to Venezuela to visit, study, and support the many Huntington's families living
along the shores of Lake Maracaibo. Drawing on these families' deep knowledge
of their histories and genealogies, Nancy and her team constructed pedigrees
that David said were "the best I've ever seen." They also took blood and skin samples and provided basic medical aid and support to the families. Eventually they
helped to open a specialized Huntington's nursing home and care center called

the Casa Hogar de Amor y Fe, which operated in the barrio of San Luis, on the outskirts of Maracaibo, for close to two decades.[2]

Dad was elated at Nancy's achievement. "I want to build her up," he told me. "And I am building her up. She deserves a lot of the credit." Part of her success in eliciting the cooperation of the Venezuelan families lay in her willingness to show them that she had a personal stake in this research: she, too, was at risk for the disease. She, too, had donated samples. They embraced *la catira* ("the blonde") as a kind of savior, one who understood their suffering because she was one of them. Just how much she was one of them was an open question that lay at the back of everyone's minds, most of all Nancy's, Dad's, and mine.

At the time of the marker discovery I was on leave from teaching at Sonoma to write about Emma Goldman, and, except for research trips, I was living with John in Riverside. At the age of forty-one, I felt time rushing by. Writing a book had been my dream ever since I was in elementary school, and once the anxiety about Huntington's began humming in the back of my mind, I did not want to waste any time. With John's generous support and a number of grants, I was well on my way to completing volume one of a Goldman biography and was already planning volume two.

As soon as I heard about the marker discovery, I knew that I wanted to write this story. All at once I felt that I had found a way to be part of the Huntington's disease community as a historian and writer. I could use my professional skills—delving into archives, interviewing the major actors, observing in laboratories and at meetings—to write about the research as it unfolded. I could place it in an historical context, weaving together a science narrative with my version of our family's experience of the disease. I loved the way François Jacob had put it: "to reconstruct worlds from which has disappeared the uncertainty of the future, a future now in the past."[3] Suddenly I had a purpose in attending all those science meetings, Huntington's conventions, foundation workshops, and fancy fundraisers besides tagging along after Nancy and Dad. I felt excited as I set out to document as much as I could before the revisions of hindsight set in. I needed to learn some science as well, since I, like Dad, had no background of any kind in biology. That my version of the family story might cause further tension with my father occurred to me, but I shoved that anxiety aside. I decided to think about it later.

The 1983 marker discovery broke new ground in human genetics, making Huntington's the first genetic disease to be mapped to a chromosome when it could have been anywhere in the genome.[4] It helped galvanize the emerging effort to map all human genes, formally initiated in 1990 as the Human Genome Project. It also provided the starting point for what became a historic collaborative effort to identify the HD gene itself. At the January 1984 foundation workshop in Santa Monica, a number of the researchers present, encouraged especially by David Housman, organized themselves into a collaborative group of gene hunters with a mutual agreement to share all materials and discoveries along the way. Afterward Dad told the participants that this was exactly the brainstorming meeting he had always dreamed of, the way the foundation workshops had been at the beginning, when they were spontaneous and freewheeling, before they had become so successful that they grew too large, in his view, and more focused on sharing data than on coming up with new ideas and experiments.[5]

The collaborative group expanded slowly, eventually consisting of principal investigators and postdocs in six academic laboratories in the United States, the United Kingdom, and Germany. Later they would agree to publish their results as a group. In the highly competitive world of biomedical research, this level of collaboration was rare at the time. They also formed an oversight committee to problem solve along the way. They decided to get together quarterly to discuss their progress in person, inventing themselves as they went, developing new technologies, new ways of communicating ("electronic mail" was just coming into being!), and new protocols for collecting samples and for sharing methods and materials. They discussed who should have access to the G-8 probe—the short single strand of DNA that attached itself to chromosome 4 in the neighborhood of the HD gene—and how they could control the way it was used.

Dad found all these discussions exhilarating for he knew he was participating in something historic. He loved attending meetings of the collaborative group when they were held in Santa Monica, and, over the course of the ten years it took to identify the HD gene, he acted as counselor, therapist, troubleshooter, mediator, mentor, ombudsman, resident sage, benevolent father figure, consigliere, and ex officio senior member. He sensed when to step in at critical moments to articulate unspoken tensions so they could be addressed. He calmed ruffled feathers. He knew how and when to open up a space for people to air grievances in a safe atmosphere. His calming presence and his expertise in group dynamics helped the collaborative group function through years of difficult work. Conversations within "the collaborative," or "the collab" as they called themselves, were as much about process as about the product they were seeking: the elusive Huntington's

gene. Establishing procedures for who got included and who did not, addressing issues of confidentiality, and defining rules for protecting junior faculty and
postdocs who needed first authorship on a published paper to achieve tenure or
get an academic job—all these issues and more had to be worked out in meetings
that could sometimes be contentious. They required delicate interventions and
emotional intelligence, areas in which my father and sister excelled.

My father also came to many meetings of the science advisory board and
even hosted some of them at his apartment. He sat in on workshops held locally,
although, as the discussions grew increasingly technical, he had trouble understanding and, at times, hearing some of the more soft-spoken participants. But
he was always enthusiastic about his one-on-one conversations with individual
scientists or with a few of them together during lunch, at coffee breaks, and at
group dinners. Over the years he became a mentor to some of the younger investigators, such as Anne B. Young, a neurologist at the University of Michigan,
a critical member of the yearly Venezuela teams and Nancy's best friend, who
turned to him for advice in 1991 when she was offered a position as (the first
woman) chair of neurology at Harvard and the Massachusetts General Hospital in Boston. Anne told Dad that she had never dreamed she would have this
opportunity, so she felt rather confused and bewildered when it came up. "Then
to complicate matters, everyone I talked to had a different opinion about what
I should do," she told him. "Your simple, straightforward and wise comments
to me . . . really helped me resolve the conflicts." Anne told me later that he had
advised her to trust her intuition ("trust the amygdala" was how she translated
it) to figure out if she truly wanted the position and what she needed in order to
accept it. If she truly did want it, then she could rationally determine how to get
what she needed. She followed his advice, successfully negotiated her conditions,
and then accepted the position and never looked back.[6]

It was a heady time, an exciting and challenging time in different ways for Dad,
Nancy, and me. In June 1983, Nancy resigned from what was then called the
National Institute of Neurological and Communicative Disorders and Stroke
to accept a faculty position in the departments of neurology and psychiatry at
the College of Physicians and Surgeons of Columbia University—George Huntington's alma mater. No longer a government employee for whom lobbying
was precluded, she now also assumed the presidency of the Hereditary Disease
Foundation while Dad moved to the role of chair of the board of trustees. He
was immensely pleased and proud, although tensions arising from the fact that

Nancy was on the East Coast and Dad and the foundation office were in California would become increasingly difficult to navigate.

Meanwhile Nancy's new partner, Herb Pardes, was appointed chair of the department of psychiatry at Columbia. They moved together from Washington, DC, to New York. Dad had gotten to know Herb during their China trip in 1981. Their shared enthusiasm for biological research in psychiatry should have formed a bond between them, especially after Herb joined the board of trustees of the Hereditary Disease Foundation and was willing to work on its behalf. But Dad had always turned a critical eye toward any boyfriend of Nancy's. And Herb, eleven years older than Nancy and married, and a powerful figure in a field close to Dad's, did not escape his scrutiny. Herb in turn came to regard Dad's influence on Nancy as excessive and sometimes expressed jealousy of their close relationship, claiming that she was more attached to her father and sister than she was to him. As Herb moved up the administrative ladder at Columbia, from chair of psychiatry to medical school dean and eventually to president and CEO of New York-Presbyterian Hospital, the flagship of one of the largest and most prestigious hospital systems in the country, Dad found himself grateful for Herb's generosity and fundraising efforts for the foundation. But he also resented Herb's claims on Nancy's time, claims that grew as Herb wanted Nancy to accompany him to the nearly nightly receptions, benefits, galas, dinners, and other fancy social events that were a part of his job. If Herb sometimes joined Nancy on her trips to Los Angeles, she would find herself torn between the demands of each of them. On one occasion, Nancy and Herb were in LA together and went to the movies one afternoon instead of visiting Dad, who was furious and blamed Herb as if Nancy had had no part in the decision, declaring he never wanted Herb in his home again, though he soon forgot this vow and peace between them returned.[7]

But all that came later. Besides advancing the search for the HD gene, the 1983 marker discovery made predictive genetic testing possible by serving as a proxy for the abnormal or normal version of the gene and hence for the presence or absence of the disease. It thereby opened up a Pandora's box of legal, social, and ethical challenges and raised many personal questions for Nancy and me. Should Nancy get tested? Should I? Nancy's scientific work had made the test possible. It seemed logical to many people that she would choose to take it. And in the immediate aftermath of the marker discovery, before testing became a reality, she expressed her eagerness to do so. I was more hesitant but willing to go along. And when the test did become available in 1986, as

FIGURE 12.2 Nancy, Milton, and me with Frank Gehry's *Little Beaver* chair of corrugated cardboard.

Portrait by Mariana Cook, 1992.

part of a research protocol, many people at risk rushed to get tested. Here was a way to "end the uncertainty" and "plan for the future," we were told. The media sometimes portrayed our choice as one between "knowledge" and "ignorance," a subtle form of coercion. But as more and more academic medical centers began offering the test, the complexities began to give many of us pause: because of possible technical errors, because of occasional ambiguous or uninformative results, because of the potential for discrimination following a positive test, because the predictive test gave no indication of *when* symptoms might appear, and most of all, because there was no way of preventing or delaying the disease.[8]

Dad had never been a fan of such tests, and soon he became adamantly opposed. Although he tried to maintain that the choice about testing was

Nancy's and mine, he couldn't help portraying testing as a collective family affair, as each of our individual choices would have repercussions for us all. Even with his calm demeanor and encouraging letters to HD families, he conveyed his dread when he described Huntington's in sensationalist terms as "the worst there is, like cancer and Alzheimer's put together." Did we really want to know? Nancy was thirty-eight. I was forty-one. There was no time to lose if we wanted to have children. Testing promised the possibility of learning we did not carry the abnormal gene and could thus have a family without worrying about passing on the disease. But what if we were not so lucky? Could we live with that? What did it mean if we decided not to have children because we feared passing on the HD gene to offspring who might not develop symptoms until their forties or later? What about Dad's statement that he would never knowingly expose a child to such a danger? Did he really believe that a life at risk or a life with Huntington's was not worth living? Ultimately the fact that there was no way to prevent, delay, or slow the disease persuaded many of us, including Nancy and me, to forgo the test. We would live with uncertainty. We both decided not to get tested. At least not then. Not yet.

Despite our rocky deliberations, the marker discovery relieved some of Dad's anxiety by convincing him that a cure might be close. After all, no one had expected the gene to be localized in just three years; finding a cure might be equally rapid. And once we came to the decision not to get tested, at least not right away, he relaxed a bit, though he remained concerned about providing for us in the future "just in case." As early as 1971, he had written to us that he was "reasonably assured" that we would not need the fund he had set up in case of illness. He had a hunch "that you and Nan are in the clear anyway," he told me. "I just want to be sure. My hunch, incidentally, is based on my own intuitive kind of clinical observation; and I'm inclined to trust that. Neither of you fit any of my observed cases at either the psychological or physical level. But that's the hunch; not proof. So I go on with my obsessive safety measures."[9]

For the rest of his life, Dad's letters tacked back and forth between certainty and uncertainty, between optimism and anxiety. He could not let go of his worry that one or both of us might develop the disease. Starting in 1977, he began organizing his finances in a formal way to address that possibility and to provide the best medical and nursing home care, because "if that didn't comfort my bones in the hereafter it would sure comfort my affections and maybe guilts in the present." He was straightforward about the fact that while he'd rather spend the money now and let the future take care of itself, he could not avoid "this damnable threat, fortunately growing dimmer every day, that makes me obsess and try to circumvent all tragic possibilities." He kept coming back to the subject, telling us repeatedly that he hoped we would put aside some insurance money just in case. And a few years later, he advised us that he was investing some funds that "will insure your care—just in case—and I no longer worry too much—but don't quite dare say it's impossible."[10]

The increasing visibility of Huntington's in the press and on television drew new friends to the foundation and to Dad. One of these was the American writer and icon Eppie Lederer, better known as the advice queen Ann Landers. Lederer was a force to be reckoned with in the world of philanthropy and in the media, a diametric opposite to the more subservient women in my father's life such as Mom and Maryline. That Eppie and Dad were both in the counseling business was another bond between them. In 1979, she published a letter in her *Chicago Sun-Times* "Ann Landers" column from a woman asking for advice about Huntington's. Eppie recommended the Hereditary Disease Foundation as a source of information and an organization to support.[11] Someone sent the column to Dad, who, with Nancy, immediately wrote to Ann Landers with information about the foundation, which Eppie also published. Soon Eppie and Dad, two champion letter writers, began a correspondence that did not cease until shortly before her death.

Ten years younger than Dad and a youthful sixty-one when they met, Lederer was at the time one of the best-known women in America. Millions of people around the world read her syndicated column daily. Offering down-to-earth and informed advice in her trademark wisecracking style, Lederer also consulted experts—physicians, psychologists, priests, CEOs, lawyers, and politicians—to weigh in on the questions she received. She had been active in Democratic Party politics in Wisconsin before she and her husband, Jules Lederer, the founder of Budget Rent a Car, moved to Chicago in 1954. This political experience had helped her create a broad network of advisers. It also gave her the confidence and chutzpah that helped her raise millions of dollars for liberal causes including gun control, legal abortion, and opposition to U.S. involvement in the Vietnam War. She counted presidents, corporate heads, movie stars, publishers, writers, and theologians among her friends. Although she started her career with fairly conservative views on issues such as same-sex attraction, she changed her mind over time, for example accepting that this was a human variation and should not be considered a pathology.

When Eppie and Dad met, she had been divorced from her husband, amicably, for several years. She was immediately taken with my father and he with her.

Soon the flattery machine went into high gear as Eppie sent letters compliment-
ing his writing and eventually almost everything about him. Soon she began
asking Dad his opinions on questions she received for her column. She also
began to press her corporate contacts for donations to the Hereditary Disease
Foundation. A single column promoting the foundation could—and did—bring
in large contributions over several years, including from estates of the deceased.
By early 1981, the letters were becoming more affectionate as Eppie enthused
about his cooking, his class, and his voice on the telephone. They would get
together, typically when Eppie came to Los Angeles to give a talk or when they
were both in New York for an award ceremony or foundation event.

If Dad was taken with Eppie and flattered by her attentions, as with Lillian
Hellman he seems also to have wished to forestall intimacies that he was not
ready to reciprocate. When he told Eppie in March 1980 that she was "wonderfully proper, wonderfully playful and wonderfully bright," he was expressing
reservations as well as enthusiasm. For all his courtly manners and what some
considered his patrician New England demeanor, Dad had little use for many
of the conventional proprieties aside from those of civility and decency. Being
"wonderfully proper" was not an asset in his eyes. Nor was Eppie's genteel style
of dressing—according to one journalist, she was "seldom seen in clothes that
would not be appropriate for a meeting with a banker."[12] His style was casual
Santa Monica, not North Shore Chicago. But Eppie was also down to earth
and direct, as well as funny. Attending her first Hereditary Disease Foundation
board of trustees meeting as a member in January 1981, she charmed scientists
and sophisticated Hollywood types alike. Though she could be impatient with
long-winded scientists, she circulated and flirted while distributing off-the-cuff
advice. ("Why aren't you and John married?" she demanded of me on several
occasions, not waiting for an answer.) Soon she and Dad were visiting every
few months, usually when Eppie had a speech to deliver or a meeting in Los
Angeles. Dad would invite her to his apartment for a homemade dinner of eggs
and onions, one of his specialties. And she was the ideal dinner guest, as she
thanked him with raves worthy of a five-star gourmet restaurant.[13] How could
Dad resist?

But resist he did. Sometime in the spring of 1981, he evidently wrote her an
apologetic letter, which she read and reread and decided deserved a serious
reply. She wanted to reassure him that there was nothing to apologize for. It
was all a bit reminiscent of Hellman. Her flirtatious response—almost all her
letters were flirtatious—suggested both her crush on him and his reluctance to
respond, in her view, out of an exaggerated admiration for her. She wanted him
to take her off the pedestal he had placed her on, as it was too high and made

Chicago Tribune

```
Monday...March 7, '88.


Dear Milton:

Here is "The Station."  Not on orignal
thought, to be sure, but it certainly
created a lot of mail when I printed it.
Everybody and his brother wanted another
copy.

It was  a delightful evening....and your
E and O  are ambrosia for the Gods. As
I said---I'd rather have that than
humming bird's eyebrow under glass. To me
it is the  finest treat in the astronom-
ical galaxy.  But only whenYOU make it.
I confess, m-mine is good but yours is
great.

I'll call you tonight or tomorrow.
Until then....Much

                    L and K,      ANN LANDERS
                       Eppie
```

FIGURE 12.3   A letter from Eppie Lederer to Milton, March 7, 1988.

her dizzy.[14] She might adore everything about him, and they certainly loved one another, but they were not going to fall in love, she declared, perhaps asking him to contradict her.

Dad was fond of Eppie. He appreciated her down-to-earth chutzpah; enjoyed her wit, humor, and admiration of him; and was deeply grateful for her support of the foundation. But he did not want a romantic entanglement. He wanted to keep a distance. Women seem to have pursued him all his life, at least so he perceived, and Eppie's letters tilted in that direction. But for intimacy, he preferred it the other way around.[15] Unfortunately I have been unable to locate most of my father's side of their correspondence. So his reply to the letter in which Eppie requested demotion remains unknown to me. But whatever he said did not cool her enthusiasm or discourage her from continuing their playful and affectionate correspondence and periodic get-togethers. The letters and visits continued, with Eppie still waxing ecstatic over Dad's eggs and onions, his chicken, his phone calls, his writing—which she thought better than Lillian Hellman's—and "our" daughter Nancy.[16]

Dad in turn boasted to her about the successes of the scientists the foundation supported and the accomplishments of his daughters. He also confided some of his dreams. "If I had it to do all over again I'd train for work in hard science," he wrote sometime in 1989, forgetting temporarily his aspirations as a dramatist. "I know that I have the kind of playful, experimental mind that could make for interesting approaches to the problems of the biological sciences."[17] At the same time, he described himself as "a kind of word monger. I like the sounds and the rhythms as much as I do the sense of the words. It's about the only thing musical in my nature." "Keep exercising," he reminded her. "Keep walking. Limit the e & o [eggs and onions] to visits with MW. Keep smiling."[18] On another occasion, Dad wrote to "Dear God," informing her that "in my lifetime, I've often enough been told where I should go to and what I should do. But I've always refused the invitation on the grounds that none of the places nor any of the activities were very appealing." He was waiting for "the WORD" from her and pleading with her to "please be kind. If you only knew how some people have instructed me in the past on where to go and what to do, I'm sure you'd be gentle and loving."[19]

Dad and Eppie continued corresponding well into the 1990s, with occasional visits along the way. Eppie brought out Dad's playful side, as when he wrote "Dear Ann Landers (alias Eppie Lederer—better and more truthfully known as Sweetie Pie)" to thank her for sending him her latest book and to proclaim that "to the Old Testament and the New Testament there has been added the Eppie Testament."[20] At the age of eighty-eight, he was still a flirt. But by 1996, the letters had become less frequent. Dad's declining health and his blindness may have had something to do with it. Or possibly their letters are simply missing from their archives. When Eppie died in June 2002, Dad was stunned. He had not known she was ill. She had been diagnosed with multiple myeloma in January 2002 and had decided against chemotherapy, which doctors informed her could lengthen her life by two years at most. Even many of her close friends had not known.

Dad did not speak or write much about Eppie after that, but he had enjoyed her company and their correspondence and appreciated the support she had given the foundation. I know he missed her.

If Dad was fond of Eppie, he was almost certainly in love with Elaine Attias, one of the few women intellectuals with whom he was intimate in his life. Sixteen years younger than Dad, Elaine had been his patient before she changed therapists and became his friend and lover. She was born in Canada and grew up in Los Angeles, the smart, stylish, and intellectual daughter of a self-made Jewish businessman with progressive politics, coming of age at a time when affluent young women were educated to become housewives. Elaine had other ideas. After graduating with a degree in economics from the University of Chicago in the 1930s, she went to work as a journalist for the International Longshoremen's Union in San Francisco, later turning to film production and producing several award-winning documentaries.[21] Eventually she married and had two children, returning to journalism on a part-time basis after they were grown, interviewing political leaders, writers, and intellectuals in many countries for feature articles and profiles. For all her wealthy Beverly Hills lifestyle, Elaine was always an activist for human rights, social justice, feminism, and the arts. I liked Elaine and found her political views closer to my own than was Dad's mainstream liberalism.

Like some of his other patients and former patients, Elaine became an ally and supporter of Dad's Huntington's work almost from the beginning. She hosted board of trustees meetings for the California CCHD chapter, and later for the foundation, at her home. She attended fundraisers, parties, and dinners and donated generously. She also did more. Sometime in the early 1970s, Elaine began traveling with Dad to CCHD meetings in New York and taking vacations with him. He began spending time at her house. But Elaine was a free spirit. Their relationship was evidently tumultuous, at least in the early days.[22] I remember a discussion they had one night when Dad and I were having dinner at her home. She argued that within a basically sound relationship, outside sexual contacts might be possible and that it was unrealistic to expect one person to satisfy all one's needs. Dad demurred and defended monogamy, saying that in his experience the "primitive" sexual instincts tend to take over. I supposed they were each talking about themselves. Later on, Dad often remarked on Elaine's "boyfriends," for whom he seemed to feel both envy and admiration.[23]

FIGURE 12.4 Elaine Attias and Milton, circa 1976.

Dad wrote loving letters to Elaine throughout the 1980s, wishing "you were here for a shmoozy Sunday" and telling her "you are much loved" when she was off interviewing intellectuals in Paris for an article.[24] With Elaine, Dad could be playful and express his fantasies and dreams, as well as give fatherly advice, as when he wrote to her, "Have courage, daring, and indifference toward your tasks," or told her he was looking forward to meeting her in Switzerland "for our Post Hotel, the gondelbaum [sic] [gondola], cowbells, and walks beside rushing streams, aphrodisiac hay and nougat ice," or said to her, "I'm filled with admiration for your get up and go. From Anchored Here in LA, also known as Milton."[25]

Dad's postcards and letters to Elaine differed dramatically from his (extant) letters to other women in his life. They were gayer, freer, and more from the heart. A card in May 1985 expressed his determination to "make sure that you have no excuse to complain about neglect. But also to wish you deep pleasure in this adventure. And for all time."[26] On April 23, 1986, he was typing a long letter late at night because "I do love you and would keep on writing indefinitely if only my eyes would stay open." But they would not, so after two single-spaced type-written pages he delivered a "big kiss and a hug and a wish that you were nearby and I could deliver in person."[27]

At some point Elaine ended her romantic relationship with Dad, finding him too jealous and possessive, as she told me one night after he died. Whether he was crushed or only mildly hurt is unclear to me, for we did not discuss it. Elaine herself was somewhat secretive—or one could say private—about her affairs. But while they were no longer lovers, she and Dad remained good friends for the rest of his life, and she, like Maryline, continued to support his Huntington's work even as she stayed involved in political causes. Intelligent, glamorous, charming, and adventurous, Elaine never seemed to have difficulty finding new companions, lovers, and friends. In 1989, Dad was still writing to "Doll," though his letters were now more newsy and less personal, and "Doll" and "Dearest" had become "Dear Elaine." But he missed her when she was away and still sent love and long letters. There is "something in your style that is remarkably attractive and inspiring," he wrote to her in a 1990 letter, after their affair had ended. "It's a simple thing. And it came to me while I listened to your plans to travel to France and Biarritz. You were not going alone. Someone travels with you and makes you happy. But you include me in and I am most happy for you and with you. I think you have these relationships with men and they are sincere and decent and loving and as they end you go on your way but manage to keep in touch, maintain friendships, add to your resources in people who love you and never seem driven to hostility or recrimination." He added that she "may think, 'Doesn't everyone?' and I would answer, 'Hardly anyone.' It's a talent.... No, you are very very special.

You have a talent for friendships. You are straightforward, open, available, kind, and unable to comprehend the petty visions of most people who end relationships only with bitterness and recrimination."[28] Dad admired those qualities in Elaine long after their romance had come to an end.

Although I was certain I knew of the women with whom my father was intimate, there was one I had never heard of until I discovered his writing about "Libby" years after he died. I do not know if she was real or imagined, but in any case the story haunts me still. Did this narrative describe the kind of relationship he had dreamed of all his life, or was it a fiction? It has the ring of truth to it, with autobiographical details that place it within a recognizable epoch of his life. The story, told in the first person, recounts a relationship of sharp intellectual disagreements and deep emotional compatibility. Libby was ten years younger than the narrator. By his own account, she mocked, scorned, and ridiculed almost everything he said. She would tell him he was a horse's ass while asking for another martini. "Now the truth of the matter is that I wouldn't have it any other way," he wrote.

> Libby was not only my best friend, she was also the seasoning that made life interesting and possible. There were times when each of us, separately or together, thought that life had played us a dirty trick. We could have been lovers or a married couple. By the time Libby became free by way of a difficult divorce, and I became free by becoming a widower after the prolonged illness of my wife, Libby and I had achieved such a marvelous friendship that neither of us wished to muck it up with romanticism, sentimentality or sexuality. It was the first time in our relationship that we had reached a solid accord without some uneasy reservations or minor caveat to ruffle the peaceful landscape.

In the story, Libby is a political cartoonist, a satirist, who had become something of a cult figure on university campuses, where she was often invited to speak. Dad admits in the story that his portrait does not sound like the basis for a good relationship. But he and Libby "found peacefulness somewhat abhorrent and dull. I think somewhere deep down, we both also found it somewhat dangerous. It was as if we might get lulled or mesmerized into some form of tenderness, opening into avenues of intimacy that both of us knew would end in the kind of somnolence that afflicted Sleeping Beauty."[29] That story had a special meaning for my father, for Theodor Reik had once, as a joke, composed his own version in the guise of a newly discovered manuscript by the brothers Grimm in which the ecstatically married prince and princess grow drowsy and somnolent, their entourage falls asleep, and the entire castle falls into ruin. My father found Reik's story amusing and accurate in terms of "the impermanence of infatuation, even of marriage."[30]

The Libby story made me sad. I asked some of Dad's friends who they thought Libby might have been in real life. None of them knew. Or perhaps they did not want to divulge a secret they thought he would not have wanted to be shared. But a fragment from one of his brother's letters sheds light on the Libby tale. "I was interested in what you said about being 'better off at some distance from real intimacy,'" wrote Henry, who could not help interpreting Dad's phrase as the psychoanalyst that he was. "If you make 'real intimacy' something like a surrender of your soul and if the necessary compromise in that state means that to you, I suppose you are right. You put it thus: if you compromise you'd feel trapped and see it as due to the other's shortcomings and this would be true enough since they fail to meet fully your needs and expectations. The other may feel similarly of course with or without the sense of entrapment." Henry surmised that for his brother, "real intimacy leads to compromise, to surrender of the free spirit, to a sense of entrapment, to anything, in short but intimacy as a sense of familiarity, closeness, affection and the like." Henry thought my father came close to thinking like Unamuno, or perhaps Machado de Assis, both of whom had drawn bitter portraits of marriage. Henry wrote that while he "almost subscribed to your thinking, I still have the feeling (or is it yearning?) that there are those who do share an intimacy that doesn't lead in the direction you speak of."[31]

One night thirty years later, Dad and I were in his bedroom after dinner, Dad seated in his chair at his white metallic desk thinking aloud about his life. He told me the one thing he regretted was that he'd never really been in love. He had read about being in love in books, he knew some people he thought were in love, he'd been attracted to people, he'd loved people, he'd been close to people. But he'd never really been in love. I sat there stunned, loath to believe my father had missed out on such a primal human experience. I thought, *In four days my father is going to turn ninety. He is feeling lonely and old, here, now, on August 20, 1998, sitting with me in his bedroom, a fine-looking old man with a full head of silver hair, dressed in his favorite navy blue polyester pajamas with the white trim.* I thought, *He no longer remembers being in love with Maryline, with Elaine, with 'Libby.' He misses love. He would like to be in love right now.*

Dad was thinking about the future. His therapy practice continued to engage him, and he loved interacting with scientists and keeping abreast of new research developments, but he wanted some distance. He was tired, he had told me after the marker discovery, tired of being beholden to those who helped the foundation, of having to express constant gratitude, of not feeling free to say what he thought. And he was tired of being considered a bully and a tyrant by his secretaries and, at times, by Nancy and me. "I could be a love, a dream, a soft saint by having no causes, no ambition," he once told Nancy. "But when I fight for what I believe in and what is important to me then I'm accused of being tyrannical. When I withdraw and do my own thing I'm accused of neglect and indifference."[32]

Most of all he wanted to write. With Nancy as foundation president, he intended to keep up his involvement in fundraising, but he wanted to be emeritus. "The foundation? It's up to you two," he had told Nancy and me the year before. "Don't let it become an anchor. Pass it on if it's burdensome. We've done something quite good. We don't have to be good forever."[33]

13

Retelling Lives

The self is always a narrative construction, always open to flux, and so never totally settled once and for all.

—Roy Schafer, *Retelling a Life*

Dad was a writer all his life, no matter what else he was doing. He devised arguments for debates, drafted legal briefs, composed scholarly papers for psychoanalytic journals, penned lectures and "Discussions" and book reviews and talks and memos and minutes and hundreds, if not thousands, of letters over the years. But he wanted to write fiction. So in 1982, when the director Blake Edwards, who was also his patient, invited him to collaborate on a scene in the movie he was filming—a remake of François Truffaut's 1977 *The Man Who Loved Women*—my father did not decline. The film is a comedy about a sculptor and serial seducer, played by Burt Reynolds, supposedly charming and irresistible to women, who is struggling to control his sex addiction. Recognizing that he needs help, he seeks out a psychoanalyst, played by Julie Andrews (Edwards's wife). For a scene in which Reynolds lies on the couch free-associating to Andrews, Edwards arranged for Dad to be hidden on the film set in a trailer from which he could feed appropriately psychoanalytic responses through a wire to Andrews, who wore an earpiece. Reynolds, who had agreed to improvise in the scene, remained uninformed until after the shoot when the joke was revealed. Dad found the experience exhilarating, Andrews not so much. "I had the peculiar sensation of art imitating life, and it felt more than a little bizarre," she wrote in her memoir, *Home Work*.[1] *The Man Who Loved Women* was almost universally panned, and Dad, too, agreed that "it didn't work." Despite the poor reviews, Edwards invited Dad to collaborate on the screenplay for a second film, *That's Life!*, about an architect in the midst of an emotional crisis as he faces his sixtieth birthday. For this film Dad received a screen credit as cowriter. Although the *Wall Street Journal* critic Peter Rainer dismissed it as "a bland swatch of knuckleheaded narcissism," it did receive some favorable reviews on account of terrific performances by Andrews and Jack Lemmon, as well as engaging cameos by several others.[2]

Such collaboration between a patient and his analyst brought Dad unwelcome criticism in the media. He sometimes claimed that he was "too old for such stuffiness" and brushed off journalists who questioned him. At other times, he insisted that he did not believe in rules against "dual relationships," rules made for "social workers." Or he justified his actions on the grounds that both Freud and Reik had formed friendships with some of their patients. But privately he acknowledged that he was "bending the rules." He told me that he knew you were not supposed to write screenplays with your patient. He was not as cavalier as he appeared to be.[3]

He continued to pen his own screenplays and a few treatments throughout the 1980s and 1990s, inspired by his experience with Edwards.[4] He enjoyed the writing process, and that was enough, even if he hoped for more, occasionally asking a well-connected patient to show one of his scripts to industry higher-ups, another cringeworthy ethical breach.[5] Edwards may have encouraged his writing, for Dad's script "Hail to the Chief," based on a 1962 treatment by the television comedy writer Nat Hiken, about a president who becomes increasingly withdrawn and irrational in the White House, bears extensive penciled commentary, most likely from Edwards, who had optioned Hiken's treatment. "Helen and the Lord," which Dad developed himself, satirized the antiscience outlook of television evangelists such as Jimmy Swaggart and his gullible followers and featured a female science professor as the hero. "Crap Shoot," another original script, dramatized the disastrous consequences of weak paternal authority within a family, one of Dad's favorite bugaboos. However, this story comes to a happy ending thanks to the tough love and clever strategies of the mother, who, unlike her spineless husband, refuses to be bullied by their son. Dad also wrote several drafts of a novella, "Penelope," a fable of disability and the creation of community, inspired, I believe, by a short story Dad had always admired for its celebration of unquenchable optimism, "The Illusionless Man and the Visionary Maid," written by his analyst friend Allen Wheelis, another ex-Tope-kan. Of all Dad's fiction, "Penelope," featuring a brilliant but extremely shy and fearful young woman with a perfect memory and great mathematical skills, best captures my father's empathy and affection for those human beings whom society so easily dismisses.[6] He also wrote a sequel in which a screenwriter comes to a California town called Mystic and discovers Penelope and her husband and cohorts living there, precisely the characters this screenwriter had invented for his most recent film. Reading "Mystic," I was reminded of what Dad had said years earlier about traveling for the first time to Rome and the "queer feeling" he experienced at finding reality almost too closely matching his fantasies. "Mystic" was in some ways Dad's most autobiographical script, with its screenwriter pro-tagonist resembling the way Dad imagined himself late in his life, as a man "who lives mostly in his head" and who "goes around creating towns and people and places [in his imagination] and every once in a rare while" runs across a locale eerily identical to his dreams.[7]

The film project closest to my father's heart was his screenplay based on Saul Bellow's novel *Henderson the Rain King* about a wealthy, white, middle-aged man from Connecticut who travels to Africa to find spiritual enlightenment. Edwards had optioned this novel, too, and written a treatment and possibly a screenplay. Published in 1959, the book reeks of familiar racist and colonialist tropes. Dad acknowledged the "defects" of the novel, but he "reverberated" to Henderson's "spiritual search for identity and for meaning," as he wrote to a cor-respondent, and felt that its emphasis on living in the moment and on human relationships highlighted what was important in the human condition.[8]

Nothing ever came of the Henderson screenplay or any of Dad's other film treatments. Fortunately he had another writing project, not a screenplay and not fiction but one in which he could use fictional techniques such as scenes and dialogue. Encouraged by Lillian Hellman and enthusiastic about his new Kaypro home computer, which he quickly mastered, he began in the early 1980s to write down memories. He enjoyed writing these reminiscences so much that he just kept going. *A Look Through the Rearview Mirror*, as he called it, was less a memoir than a series of vignettes; a portrait gallery of the vivid characters he had encountered during his life and a self-portrait through the lives of others—from his Brooklyn high school teacher with narcolepsy who fell asleep for brief naps during class; to the con artist Leo Chertok who had cajoled Dad, as a neo-phyte lawyer, into defending his alleged rights to oil in Ethiopia; to Dad's analyst, Theodor Reik, with his photographic memory and didactic stories; to a lonely, little-boy genius who developed a clever and heartbreaking ruse to pretend that he had friends; to the mercurial and mysterious Karl Menninger and the flamboyant John Rosen; to "Jimmy," in reality Billy Cook, the infamous "Hitchhike Murderer" who had shot dead six people in cold blood and left Dad, as an expert witness in Cook's murder trial in Oklahoma, with the sense of coming "as close to the essence of evil as I'd ever care to confront."[9]

Dad did not want to probe his inner life or explore his intimate relationships in a published memoir, in part, I believe, because of the pain of reliving parts of his past and because of his strong sense of privacy.[10] Would he have approved of this biographical memoir, I ask myself, with its deeply personal revelations? He had been enthusiastic when I had proposed writing his biography in 1969; he thought we might get to know each other better and immediately recorded a tape of memories and thoughts leading up to his arrival in Topeka. He set no limits on the project at that time, although we did not pursue it far enough for the more contentious issues to arise. Fourteen years later he was more conscious of his status as a public personality, an authority figure, the head of a foundation. He wanted to tell stories, to entertain and instruct. He wanted to have fun.

Privately, though, he continued to question his life, as if still searching for a larger narrative that felt true to his experience. He had many soul-searching discussions with his younger friend Arthur Golding, a retired vascular surgeon, also from Brooklyn, whom he regarded as thoughtful and philosophically minded despite their political differences. A memo he once wrote "for myself," even if addressed to Arthur and his wife, Dewanna, posed questions he would return to again and again: "What were we after last night? Not the definition of how to live but rather what to be." Put another way, "Who are we? What are we? How to live comes after that." That central concern of his was always present, though it seems more self-reflection than social observation: "We don't act according to our wishes but according to the sensed wishes and needs of others."[11]

Dad did not exclude himself from the criticisms he made of others. Especially in those late-night conversations with Nancy and me that we had around his milestone birthdays. One night shortly before he turned ninety, he launched into an extended meditation on all the mistakes he had made in his life, how impulsive he had been, how he had just said yes to anything interesting that came his way, from job offers to a position in the Navy to buying a summer cabin and a boat. He had been spread too thin, his knowledge was broad but not deep, even in the realm of psychoanalysis. He described himself as "the ultimate dilettante."[12]

The year before he died, Dad spoke about how he felt that much of his life had been wasted, that he had made too many "left turns," that things had been too easy for him and he had coasted instead of pursuing what he really wanted, such as going to the Yale School of Drama and writing plays.[13] I interpreted these self-criticisms more as regrets of that specific moment than as statements of a deeply felt state of mind. I was certain that my father cherished the life he had led and the people he had loved and continued to love, that he felt pride in his achievements. But I imagined that he also pondered what else he might have accomplished in a less pressured life, without the ever-present anxiety of Huntington's and with time for one of the "pure think jobs" offered to him, or if he had been able to take a sabbatical for a month or a year to write the "Autobiography of Ideas" he had once contemplated, and to interact with intellectuals outside psychoanalysis, as when Robert Hutchins invited him to the Fund for the Republic in Santa Barbara for a year. His late-night reflections moved me as expressions of his continuing self-analysis, his inner dialogue, which is what he said psychoanalysis turns into after the dialogue with one's analyst comes to an end. Though he boasted at times, about the foundation especially, he also expressed a surprising humility: "the most ambivalent attitude toward achievement of any high achiever that I've ever known," according to one acquaintance.[14]

By 1990, Dad had accumulated more than five hundred pages of stories from his past. He did not like to revise, much to the frustration of his writer friends. And Nancy and I were occupied elsewhere. His manuscript sat in the foundation office until early 2002, when we decided it was time. After two literary agents deemed it too anecdotal and "too episodic for a book," we decided to publish it ourselves, with editorial help from my friend Ellen Krout-Hasegawa, an astute and imaginative reviewer and critic working at the *LA Weekly*. Dad was in his nineties and growing frail. We wanted him to see it in print and have a book party to celebrate his life before it was too late. And so we did, at Elaine Attias's home, with a grand mix of my father's friends and some former patients, as well as my friends and Nancy's. We lined up the chairs in the living room theater style with Dad seated beside the fireplace, the aide he called Singing Tess near him just in case. Nancy and I read humorous sections aloud, and Dad and Carol Burnett made everyone laugh with their repartee. We had fun. And so we launched *A Look Through the Rearview Mirror* into the world, where some people read it and liked it, and that was that.

I had long pressed Dad to distill into a book the psychological wisdom that he had gleaned from more than fifty years of doing psychotherapy—not so much his psychoanalytic wisdom but his insights into how to live a meaningful and rewarding life. Like a jealous child, I imagined that he had shared more of these

insights with his patients than with Nancy and me, his own daughters. And that
he would go to his grave without revealing to us the full extent of his knowledge.
Eventually he decided to take on the project, thinking that such a book, if it sold,
could make money for Huntington's research.

To frame the advice in the book, my father conjured up two contrasting role
models. One was Benjamin Franklin with his procedures for improving his
character and acquiring moral virtues. The other was the French artist Marcel
Duchamp, who allegedly gave up making art for playing chess when he had the
epiphany, "I am the art." This story, apocryphal or not, captured Dad's imagination as a way of thinking about shaping one's life.[15] Despite his conviction that
"character is destiny," Dad argued that people could alter their character in more
aesthetic and satisfying directions; he imagined himself as a sculptor helping others mold their lives into works of art.

He did not pretend to offer great revelations or deep insights; he knew he
was writing pop psychology based largely on his experiences in late twentieth
century Los Angeles and with people who had options and opportunities most
did not have. Much of what he wrote was familiar. He pleaded for the acceptance of different emotional styles in relationships, extrovert or introvert,
effusive or reserved, barnacle or porcupine; in his view each had its validity.
He encouraged the supremely American willingness to take risks and not be
defeated by failures, having the courage to pursue one's vision, whether in a
job, in art, in architecture, or in something else. Most of all he insisted on
the value of freeing oneself from dependence upon the approval and attention of others and of developing one's own way of being in the world. He
cherished the individuality of artists such as Frank Gehry, Quincy Jones, and
John Altoon, not the greedy individualism of a Donald Trump and his ilk.[16]
Though writing in a post-Reagan world, Dad at ninety still assumed a basic
social solidarity, a "New Deal" vision of society. He believed there were values that all intelligent people might agree upon, something akin to the Ten
Commandments, but that these did not define us. He disliked such concepts
as duty, obligation, obedience, and conformity to the expectations of others.
He championed courage, commitment, generosity, responsibility, and loyalty,
which he saw as values one might freely choose. "Winning Ways" offered no
answers but urged readers to "practice aloneness for a minute, for an hour, a
day, a week or a month. Practice becoming your own self-validator. Take your
life back." Instead of rushing off to a new mother or father figure, that is, a
therapist, "to squirt some courage into your veins," he encouraged readers to
try thinking on their own. He was certain that "a little thought on your own
may go a very long way."[17]

But for all his emphasis on self-validation, he also accorded great importance to thinking *with* others. He valued dialogue, conversation, discussion, and exchange. This was the Milton Wexler who had once confided to David Rapaport, "I am afraid I do not really work well alone," the Milton Wexler who needed to be "pressed by outside forces in order to be productive," who valued people "who would push me to more creative work," who needed interlocutors "as the cuttle-bone to sharpen my own wits." Dad worked best in dialogue with others. "Winning Ways" included a long encomium to Franklin's Junto club, a weekly discussion group of men committed to following certain conversational rules—avoiding ridiculing and dismissing others, for example—in searching together for truth. Members believed that they would become more skilled at dealing with people and situations by participating in the group's meetings, which my father viewed as an early form of group therapy. Dad was also an early advocate of support groups and believed Alcoholics Anonymous was far more effective in addressing alcoholism than individual psychotherapy. "Winning Ways" even proposed extending the Duchampian metaphor: "*We* are the art." Starting out to write a treatise on shaping oneself, he ended up making a case for social interconnectedness and mutual aid.

Like most self-help writers, Dad was most concerned with the ways that people sabotaged themselves, not with how large social and cultural forces such as racism, misogyny, poverty, and war created enduring psychological trauma and destroyed people's lives. "Winning Ways" addressed emotional suffering primarily as a problem of the individual. But he acknowledged that for many people, life was harsh, brutal, and unrewarding through no fault of their own. While praising stoicism in the face of pain and the refusal to burden others with complaints, he also honored "righteous rage on behalf of your fellow creatures, on behalf of all living organisms, on behalf of the planet," which he thought "may represent an important aspect of true spirituality." He emphasized that while a good-humored approach to life is likely to bring better outcomes, real tragedies "should not be faced with good humor. Grief, anger, [and] painful psychological work may be the order of the day."[18]

And where was Freud in all this advice or the Freudian human being driven by aggression and lust? I sometimes wondered whether my father was still a psychoanalyst or had become a life coach. By the 1970s, he had come to believe that classic psychoanalysis was appropriate only for a small number of people with the motivation and resources of wealth and time to pursue an extended process of self-exploration. It was not a preferred method for overcoming emotional conflicts or altering behavior, as even Freud acknowledged. Insight alone did not bring change. Dad thought psychoanalytically informed psychotherapy had much broader application than psychoanalysis. Late in his life he

FIGURE 13.1   Milton with Frank Gehry, circa 1980s.
_____
Photo by Elaine Attias. Courtesy of Dan Attias.

sometimes claimed that people could learn as much from reading the novels of
Trollope or Dostoyevsky as from seeing a therapist, although I don't think he
really believed that.

What he always believed in was the possibility of change. "While character tends to get fixed at an early age," he wrote, "the idea that it is forever after
immutable and unchangeable is ridiculous. There are so many evidences of transformations, of growth and of development. There should be no room for despair."[19]

Clashes were erupting with growing frequency between Dad and me as he felt his
life, at moments, starting to drift. Writing to Elaine in the fall of 1989, he alluded
to "the boring nature of my life," wondering what he did all day. "I get up, see a
few patients, take walks, do some writing, go to a movie on rare occasions, and
take medicines. I try to keep my good humor, erase all kinds of vain regrets, and
stare at the ocean and the palm trees. Not even a good book to read."[20] He was
working on his novel and on a screenplay, but these projects were not enough
to fill his days.

Meanwhile I was sinking into depression, sitting alone in my study at home in a silent Riverside suburb or in the university library, day after day, working on the book that became *Mapping Fate*. As I look back at my diaries of the time, I am surprised to see that doing what I wanted to do—writing a book to which I was deeply committed—and achievements that should have made me feel happy offered only temporary relief: a fellowship, a contract for the Huntington's book with a sizable advance, and the publication of a second Emma Goldman volume that led to some invitations to speak. But comparing myself with my father and sister who were out in the world doing important work, I struggled to define my professional identity. Was I primarily a writer? A professor? Most professors were also writers. Why couldn't I do both? Why did I feel like a failure? In the fall of 1990, Sonoma notified me that I would either have to return to teaching or retire; I could not stay on leave forever. I chose to retire, knowing that I was fortunate to have financial support from John and from my father, as well as income from grants and from the classes that I taught part-time at several institutions. The thought of teaching full time again made my heart sink. Thinking of myself as a writer felt more congenial. Yet even with my financial privilege, the identity of a writer was hard to sustain while we were living in Riverside, where almost everyone we knew was associated with the university, seemingly the only site of local intellectual and feminist life. I felt unmoored, a permanent outsider. My depression was starting to spread to John.

I also felt uneasy about the book I was writing. "I knew in advance I would love parts of the manuscript and hate others," Dad told me frankly in 1986 when I showed him a draft of the introduction. But he encouraged me to write the story as I saw it. He knew Maryline would also be upset but urged me to tell her I was writing from inside my experience, not as an outsider. "How will it be with my patients? Pretty awful, I suspect," he wrote. But he did not object; at least he tried not to. "For some time now I've been freeing myself from accountability to anyone," he told me. "That's hard, especially in light of the Foundation. But there's a fresh breeze blowing through my head and it says I don't have to explain myself, account for myself, answer to anyone. To some extent, in a strange way, it feels like being a grown-up for the first time."[21] But a few years later, Dad changed his mind, claiming at moments that the book was really about attacking him.[22] His new view disturbed me greatly because I was beginning to feel that the book was working. What had begun for me in a burst of enthusiasm, as an effort to bear witness to a critical moment in the history of genetics and of a genetic disease, felt increasingly like disloyalty, even treason.

Late in 1990, I was feeling so depressed that I decided to seek my father's advice. By this time in my life I did not ordinarily share intimate thoughts with him, but at the very least he could recommend a good therapist. Reluctantly and with considerable anxiety, I decided to tell him about my depression, something I had never really done before. One day in January 1991, I drove to Los Angeles to have a talk with him over dinner. I began telling him how low I was feeling and how puzzled I was that I could not seem to escape my depressed thoughts. He listened patiently. All of a sudden he burst out, "Annie has AIDS!" Along with Edy, Annie was a longtime trusted secretary of both his therapy practice and the foundation. He explained that Annie had had a brief affair with a man who had not informed her that he had the virus, had AIDS in fact. She had just received her diagnosis a few days earlier. Dad was devastated.

I was deeply saddened but also angry; I felt completely silenced and dismissed. What more could I say about feeling depressed when Annie was going to die? Furthermore, as he told me, she had asked him to keep the information confidential, and he had just violated her trust. Then, inferring that foundation activities were contributing to my unhappiness, he suggested that after I finished my book I should feel free to distance myself from the foundation since it did not seem satisfying or meaningful to me. I did not say much for the rest of the evening and let my father talk, mostly about Annie. But before leaving his apartment early the next morning, I wrote a note telling him how sorry I was about Annie but also how upset I felt at what I perceived as his obliteration of my concerns. I told him that it did not seem fair for him to tell me, just at that moment, that Annie had AIDS and had not long to live. I also told him that his suggestion that I distance myself from the foundation (advice he had also given to Nancy at one point, though I did not know it at the time) felt like a rejection and another instance of his dismissal of my actual concerns, concerns I believed had little to do with the foundation. I felt as if he had not heard me at all.

A few days later I received a three-page, single-spaced, typed response, one of Dad's masterpieces of outrage. He did not get angry often, but when he did he was formidable, marshaling his skills as a debater, lawyer, and psychoanalyst to make his case. I felt like a Lilliputian in the land of the Brobdingnagians. (More than once, Nancy called me to advise that I phone Dad right away if I wanted to salvage my relationship with him). "Dear Alice," he began, "I am beyond distressed. I am angry. You insist on making everything I say and do into some ugly mud pie and tossing it into my face. You have some hidden agenda which makes it impossible for you to see in me any other motive, value, or intent except something trivial, insincere, unfaithful or unkind. It bends your perceptions of me toward the dark end of the spectrum of human behavior."

Whereas he had intended to give me permission to distance myself from the foundation, to give me his blessing to go my own way, "you rapidly turn around and suggest that I am dismissing you from Nancy's and my terrain." He thought he was telling me to "go and be happy. For which I get duly clobbered as usual." He had told me about Annie, he said, "because it becomes more and more painful that you are so outside my personal orbit, so distant, so critical, that there are things I can't share with you. The reaction is always the same, judgmental, harsh, skeptical. And you can't excuse yourself by saying that everyone knows that this is in your character and not isolated behavior with just me. It still is bothering me and makes for more and more distance. I'm very glad, in fact, that much of my personal life is not in your caring hands."

He had told me about Annie because he had wanted me to share in a concern that touched all of us, and "I hate to see you excluded. So I get carried away and include you. Like pretending we're a real family and share all things equally. I'll be more careful next time. I'll put up the barriers even further, knowing all the time that I'll get clobbered once again for the distance and the barriers." He went on to take issue with my inclusion of Maryline as his lover in my book and my appeal to "historical verity as the obligation of an historian," which he thought was "bullshit." "If you are not hopeful about a better relationship, I join you in that despair," he wrote. "I've tried to keep my temper in check for many years. Unfortunately I have finally lost it. . . . Frankly I don't want to dance to your music in life. It screeches and may be very modern and creative but it seems to my conservative and elderly soul not only atonal but dismal. Sorry to be so direct and undoubtedly assaultive myself but it's time to give you my most immediate response, even if unwelcome."[23]

That was Dad in high dudgeon and it hurt. But I saved the letter and reread it from time to time, trying to feel what he may have felt. Had I thought more deeply about what he was struggling with at that time in his life—his lost vision, his loneliness, and in general "the boring nature of my life"—I might have been kinder and more generous toward his grief about Annie. I might have understood that his explosive attack was more about his own sorrow, his own feelings of loss, than about me, though some of his critique was undoubtedly justified. Dad had no grandchildren at a time in his life when they help buffer the losses of aging, when so many older people make grandkids the emotional center of their lives. I am certain that he had to listen to endless stories from his patients and friends about their remarkable grandchildren. But he never complained or showed self-pity. Perhaps Annie had been a kind of granddaughter, or a substitute for one. And now she was seriously ill and probably would not survive (she did not). Had I thought more about these things, I might have been less hurt and angry.

I might have realized that his grief at the moment simply overwhelmed his ability to listen to me. I would have known that he had sought comfort from me just as I had from him. Had I known then that he and Nancy had had similar conflicts, that he could be quick to feel attacked by her and she to feel dismissed and not listened to, just as I felt, I might never have written my note to him at all. As happened over and over, a few weeks later we had both moved on, and the argument was abandoned if not forgotten.

I was beginning to have a manuscript I liked, a new experience in my writing life. But while all of us—Dad, Nancy, and me—thought my book could help expand awareness of Huntington's, we differed in how much our intimate lives were relevant to the story. I believed that I could not write honestly about our experience of Huntington's, or about myself, unless I was specific, that the more intimate and true to my life I could be, the more my narrative would resonate with readers. This meant including Dad's affair, since surely it had contributed to Mom's depression and withdrawal and to my own adolescent confusion. It was part of *my* life, as well as theirs. But Nancy, Dad, and Maryline did not see it that way. They raised legitimate questions: Did my way of defining my project give me the right to invade the privacy of others? How would I have felt if Dad, Nancy, or Maryline wished to reveal parts of my life that I considered private and "nobody's business," as Maryline considered hers? How does one draw the line between respecting privacy and colluding in secrecy? Who should decide?

Eventually, despite his discomfort, my father agreed to accept my wishes. He was willing to overcome his fear that my narrative would have a diminishing impact on him so that I could tell our family story—and my own story—as I had experienced it.[24] That courageous and loving decision lifted a weight from my shoulders and helped to ease my depression, as did the therapist I began seeing and the antidepressant medication that I took for a time. Eventually *Mapping Fate* had a healing impact on all of us. My father learned that readers did not think him a "monster," as he had feared, but quite the opposite: they admired his commitment to taking care of Mom after their divorce and through her long illness. They admired his lack of self-pity. Most of all they admired his fierce, unwavering dedication, over decades, to finding a cure or treatment for Huntington's disease, the great mission of his life.

Ironically, someone who evidently had not read the book remarked years later, apropos of that dedication, "Your father must have loved your mother very much." "No," I replied. "He didn't love her. He loved his daughters. He did it for us."

14

Life Underwater

Not everything that is faced can be changed, but nothing can be changed unless it is faced.

—James Baldwin, *The Fire Next Time*

The earliest memory I have of my father not in control of himself or the situation was when I was thirty and he, Nancy, Maryline, John, and I spent a few days aboard a rented houseboat on a murky little body of water in Northern California called Clear Lake. It was the summer of 1972, and I was about to begin my job as an assistant professor at Sonoma. Dad had decided that floating around for a few days on this lake, which he'd never seen, would be a fun family vacation. (He'd sold the Lake Tahoe cabin years earlier.) One late afternoon we were out on the lake when the wind picked up and the current grew stronger. Dad, a strong swimmer, had been swimming near the boat when he saw it drifting away from him. He had to swim hard, harder than he'd ever had to swim in his life, he said later, to reach it again. When he finally climbed onto the deck he was shaken and out of breath. He told us he had not been certain he could make it. I had never before seen my father like that.

Dad had been lithe and athletic from his youth and was rarely sick. So when his body became less dependable as he grew older, he found himself in extremely frustrating situations. Doctors are said to be the worst patients, and while Dad was a PhD, not an MD, he was no exception. In the mid-1970s, he had a biopsy on his prostate that showed no malignancy but led to an infection that landed him back in the hospital with an alarming fever and drop in blood pressure. He came close to dying. By the time I got to Cedars-Sinai to see him, he had been transferred out of the ICU to a regular room and had perked up to the point that he was at war with his doctor, Bernie Salick. I began seeing yet another side of my father, a man unexpectedly fierce and often funny under pressure. Evidently when several inexperienced young interns came into his hospital room to examine him, Dad became so offended by their abrupt manner that he refused to cooperate, provoking their ire. "You know you are in the hospital to be treated by doctors," one of them dared to say to him, whereupon Dad flew into a rage and threw them out of the room. He then told Bernie that they should never be allowed to become physicians because they had zero psychological sensitivity, and why was he on a teaching floor anyway? Later Bernie told me, "Your father wants to be omnipotent, and he can't stand to have someone else in charge. He wants to control everything." "Oh, Bernie," Dad scoffed to me later. "He wants to be a big man." Dad was not given to self-dramatizing, but he deemed his experience in the hospital "horrendous . . . the worst thing I've been through in my life." He who helped heal so many now found himself in the role of patient, and he did not like it. His near-death experience seemed to bring out both an uncharacteristic imperiousness and unexpected comic gifts. Or perhaps I was just viewing my father in a new light. When I saw him at his apartment the following month, he recited to me verbatim his dialogues with the hapless Bernie, dialogues which made me laugh in spite of myself. " 'The trouble with you [Bernie] is that you expect your patients to lie there and not say anything,' " Dad recounted. " 'Whereas I have a brain in my head and ask a few questions. And you're not used to that.' " Dad told me that every checkup with Bernie brought with it the possibility of some awful catastrophe like a probable cancer. My father the psychoanalyst believed that Bernie wanted to come near killing him so he could then rescue him. He was going to find a new doctor.[1]

As it happened, he had few health problems after that for more than a decade, despite a diet short on green vegetables and long on hot dogs and ice cream. Then in early 1988, when he was spending a weekend in Palm Springs at Max Palevsky's, he fell asleep as he was talking to another houseguest. After a few hours he felt alert enough to drive the guest back to her home in Los Angeles. But the next night he wrote to me from his bed saying he realized that he had not fallen asleep but had passed out from heat stroke. The occasion had reminded him of his mortality. He wanted me to know that while he told everyone in the world "what a wonderful human being you are," he never quite said it "straight and direct" to me. "I think you are a spectacular woman," he wrote to me, "sincere, intelligent, dignified, interesting, creative, with admirable values and a strong sense of self that knocks the socks off of most of the people I know." These were not the qualities Dad admired most— he preferred warmth, generosity, and a great sense of humor. But still I was deeply moved. These were beautiful, loving words coming from my father. He acknowledged that "having a small and apparently unimportant crisis makes one aware of a vulnerability that belies the crazy sense of immortality that ordinarily operates in my deluded mind."[2]

Dad's concerns were not unwarranted. Three months later he was home alone reading when he began to feel weak. Suddenly he had difficulty speaking. The episode passed, but he was concerned, and the next morning he called John Mazziotta, a neurologist at UCLA who had served on the foundation's scientific advisory board. John advised him to go to the hospital. His physician, Keith Agre, came in to see him, and while he was there, Dad suffered a stroke right in front of Agre. By the next morning he was entirely paralyzed on his left side, and his left arm felt so heavy that he had to lift it up with his right. At moments he suffered leg cramps so severe that he screamed in pain, yet he demanded to get out of "the UCLA jailhouse." He hated institutions, he insisted; if he had to remain there he would die.

Dad soon made peace with the fact that if he was to regain movement and strength, he needed immediate physical therapy and a period of intensive rehabilitation. Once he accepted that, he applied himself to the job without complaint. Within a day or two, he could wiggle the fingers of his left hand; it was the first time since the stroke that he had moved any part of his body on his left side. "That's a beautiful sight," I said to him as tears came to my eyes. "Alice," he said sternly, "I've faced much more difficult things in my life." And indeed, over the next several weeks I again saw my father in a new light, at his most vulnerable and his most courageous. He took to rehab as to a serious project, focused and determined to get through it, no matter how exhausting and frustrating it was. Though he got depressed and the continuing cramps in his leg sometimes made him cry, he always managed to regain his composure, remaining mostly cheerful and charming to all the nurses and with us. He was funny, too. He told jokes. When he was younger, he had always seemed to stand in the shadow of Romi, who was a natural performer. But now Dad was the star. One day he told me that he thought it best to keep a positive attitude and that by being pleasant he got better treatment. And in fact the nurses awarded him the title of "Super Patient" and "Best Patient" for the week. He was starting to sound more like Norman Vincent Peale than a follower of Freud. When Arthur Golding came to visit and occasionally found him lethargic and unresponsive, Arthur would lean over the bed and shout into his ear, "Milton!

If you don't wake up, I'm going to register you as a Republican!" That threat always brought Dad back to life.

After ten days in rehab we brought him home to his apartment in Santa Monica, where he remained irrepressible, repeating his mantra about how it was good to have a positive attitude, which made things much easier. Gradually he recovered his strength, with a lot of help from Nancy and me and a veritable battalion of doctors and physical therapists; an exceptional nurse practitioner named Anne McKenna; his loyal assistant, Edy; and Ruth, the psychotherapist daughter of Bruno Bettelheim, a woman my age whom Dad had known through her father for many years. Ruth showed up at the hospital to help with errands and stayed to become Dad's companion for the next ten years.

Two weeks after Dad came home, John and I got married. The summer before, we had taken a glorious road trip to see *The Lightning Field*, a luminous work of land art by the sculptor Walter de Maria in Southwest New Mexico. Afterward, we drove north to Santa Fe and Taos, and, while having a drink at the historic Taos Inn, John asked me to marry him. For all my qualms about marriage and becoming a wife, I loved John and wanted to be with him. We were coming to terms with the fact that we were not going to have biological children of our own. We had tried to conceive once the Huntington's predictive test became available, making it possible, in theory, to have a child without the risk of passing on the abnormal gene. But after pursuing two years of intensive fertility treatment, we had not succeeded. We decided that even if we had no children, we could at least get married, which felt like claiming our place in the world as grown-ups, as a family, even if we were just two.

That September, three months after his stroke, we celebrated Dad's eightieth birthday and the twentieth anniversary of the foundation at the unfinished "Binoculars Building" in Santa Monica, soon to be the headquarters of the advertising agency Chiat/Day. Jay Chiat was an adventurous client of Frank Gehry, who was renovating an old concrete warehouse on Main Street, along with the artists Claes Oldenburg and Coosje van Bruggen, creators of an enormous steel-gray sculpture of binoculars that served as the entrance to the building's parking garage. Although the renovation of the Binoculars Building was still in progress, Frank's iconic wood-frame fish, which constituted the building's conference room, had been completed and, for the party, was decorated with huge black-and-white photographs of Dad at various stages of his life. Such an A-list group of celebrities wandered through the gigantic fish and out onto the patios—friends and foundation supporters, patients and nonpatients alike— that the society columnist of the *Los Angeles Times*, Marylouise Oates, devoted an entire column to the party. "In his thirty years of Southern California practice,

Click

‹ Wednesday, September 28, 1988. Los

Analyzing the analyst

Richard Rouilard

"My mother died of Huntington's. And there's a likelihood that I've inherited the disease. You're not beyond the possibility till your 60s or 70s." When she was tested a while back in Chicago, the diagnostician advising her there was yet no sign, nonetheless said the only thing to do was "not get pregnant." Miss Wexler laughed. "I just work harder at the foundation."

"I went to two psychiatrists before him," said Sally Kellerman as she bit into her hot dog at the Chiat/Day offices. "When I first went to Dr. Wexler, all I did was sob all day. He said, 'Sally, stop sobbing, it's ruining your life.' I don't know quite how — I stopped sobbing. My life changed."

Yes, Dr. Milton Wexler changes lives, the lives of the rich and famous, the not-so-famous, and — through the work of his Hereditary Disease Foundation — thousands who suffer from a panoply of diseases that his foundation is investigating.

And hundreds showed up at the exotically designed Chiat/Day advertising offices in Venice to

Everyone there had not a good, but a great word, for the doctor. Some were old friends, new friends, patients — Elaine May, Patti Reagan Davis, Jackson Browne, artists Ed Moses and Bob Graham, Quincy Jones, Fred Weisman, Henry and Ginny Mancini, Jay Chiat, Rebecca Marder and Norman Lear.

Said Dr. Wexler, when told of all the huzzah-ing: "I'm not such a great analyst. They all exaggerate, and they obviously still need help."

WHILE EVERYONE ELSE WAS out saving the whales, saving the tigers or saving the queen, Gale Hayman was letting the leopard save her

Dr. Milton Wexler got the treatment from Julie Andrews, left, Carol Burnett and Sally Kellerman — they sa

FIGURE 14.1 Milton's eightieth birthday party, 1988. Left to right: Julie Andrews, Milton, Carol Burnett, and Sally Kellerman.

Los Angeles Examiner, September 28, 1988.

Wexler has managed to become an integral part of his patients' lives," Oates noted, a double-edged remark to be sure. But she acknowledged that "Wexler is legendary for fueling the creativity of his patients and friends," commenting that one frequently becomes the other.[3]

Over the next decade, my father's medical setbacks alternated with Nancy's successes. In late 1988, the Nobel laureate and co-discoverer of the double-helix structure of DNA, James Watson, invited her to head the prestigious Ethical, Legal, and Social Issues (ELSI) Working Group of the Human Genome Project, of which he was now the director. With a mandate to support research aimed at fostering the equitable social impact of vast new stores of genetic information, the ELSI group, active from 1989 until 1997, brought Nancy into contact with the most influential social science and humanities scholars and activists involved in such discussions worldwide and with many of the laboratory scientists working in the field.

Meanwhile this new information flowed into a social world marked by grow-ing inequality. By the late 1980s, we were entering what the political scientist David Callahan has called a new gilded age, with wealth flowing increasingly to a tiny upper swath of society as the middle class shrunk and poverty expanded. Corporate salaries ballooned while wages stagnated. Democrats as well as Republicans were dismantling the social safety net, getting rid of consumer pro-tections and shrinking the public sector in the disastrous ways that have come to haunt us. Privatization of the economy and worship of the market became the new orthodoxy, and philanthropy assumed an increasingly influential role in funding science and in meeting social needs. In this setting contributions to the foundation swelled. Dad now proposed building an endowment to ensure that the foundation could continue to fund grants and fellowships and hold work-shops, a proposal to which the board of trustees agreed. So in 1985, they began transferring a portion of the funds raised each year into a science endowment that could not be tapped for administrative costs. By 1989, the endowment had reached nearly $5 million, and the foundation began aiming for $10 million.[4] One of the foundation's most memorable fundraising events took place in New York at the World Financial Center on November 8, 1989, the night before thousands of euphoric East Berliners crossed the Berlin Wall, which for decades had sealed off East Berlin from the West. From our hotel rooms in Manhattan, Dad and I watched on television as the beginning of the end of the Cold War unfolded on the screen before our astonished eyes.

Perhaps it was the contrast between the historic events transpiring a conti-nent away and our high-society activities in New York that unsettled both Dad and me. He found the New York world of fundraising oppressive. A week after the foundation gala, he wrote to Elaine Attias that he was "disgusted with the shenanigans of such an event and mighty relieved to have it over with. I'm still for non-event fundraising," he added. "There is something corrupt in all that manipulation of people. I hate it."[5] Still, he was grateful for the contributions and ecstatic to see the attention accorded to Huntington's disease in the media, especially to Nancy, who had "reached the stratosphere," he told Elaine. "Nancy is becoming a special kind of star," he boasted, "getting accolades from scientists and all kinds of laypeople, including the Cardinal of New York." He added, "It's interesting that as her dreams get realized she gets more modest, rather than less."[6]

But as the foundation's resources increased and Nancy's prominence soared, Dad worried that she had taken on too many responsibilities. "Even you cannot, like Atlas, support the world," he told her. He no longer felt able or inclined to run the foundation and was determined "to salvage what I can of the time ahead of me, for my own purposes or my own pleasures." He felt confident that the foundation was supporting good science, but he worried about the "financial end and the image making." Despite his desire for continuity, he wanted Nancy to know that if the foundation folded, so be it. He would not hold her responsible. He wrote, "And I, given back my life and my freedom, will bless and love you forever and consider the Foundation a small sideshow that has energized us both over the years and may or may not flower forever."[7]

In the spring of 1990, just two years after his stroke, Dad began developing wet macular degeneration (leaky blood vessels under the retina), which threatened to render him legally blind. He underwent two sessions of laser therapy at the Jules Stein Eye Institute at UCLA. But while the therapy may have slowed the advance of the condition, he believed the lasering inflicted its own damage. Dad felt his vision worsen after each session. In *Rearview Mirror*, he wrote that his greatest fear was that he would no longer be able to see his daughters' faces and he described his relief when he discovered that with his remaining peripheral vision he would still be able to recognize us. At that point, he wrote, "I began to push out the self-pity." Once again I saw my father's courage and determination as he confronted his encroaching blindness. As he had two decades earlier, he turned to the story of the flood and how the chief rabbis had taught the Jews to live underwater. "And that's what I'm going to do," Dad told Nancy and me. "Learn to live underwater."[8]

With his usual resourcefulness, he began to focus on becoming even more organized than he already was. He simplified his surroundings, made grocery lists in large print so that he could shop with a housekeeper, and wrote out big schedules of his patient appointments. He memorized where he put things so he could access what he needed. He had been dictating letters as far back as Topeka, so he was comfortable with that mode of writing. He gradually became accustomed to listening to books on tape. Like many men of his generation, though, he resisted using visible aids for his disability, such as the white cane we got for him. At times he tried to pass for being sighted. Once I took Dad, without his cane, to an exhibition of work by the artist Alejandro Gehry, Frank Gehry's son. When we arrived at the gallery I discovered that the paintings were enormous close-ups of young women kissing and having sex. They were brilliantly colored and in your face. Dad asked me to tell him what they portrayed, so as we walked arm in arm through the gallery rooms I tried my best, though I found describing explicit sexual activity to my father unnerving. But I figured my father the shrink had seen and heard everything, and indeed he took it all in stride. And when we finally managed to make our way through the crowd to talk to Alejandro, Dad praised the paintings with gusto. I was not sure how much he had really been able to see, but he was enthusiastic and not at all embarrassed, and so neither was I.

While Dad was dealing with the lingering aftereffects of his stroke and his vision loss, I faced a crisis that would shake my world and challenge his. John had a sabbatical for the summer and fall of 1991, and we were planning to spend it in New York. Before we left Riverside, however, things took a turn with serious repercussions for John and me and challenged my father's psychoanalytic beliefs. I began to realize that the friendship that we had with a new faculty member at the university had, for me, a sexual charge that I had not experienced with a woman before. Not only women's studies but also gay, lesbian, and queer studies had exploded on campuses across the country. Even the relatively conservative campus of the University of California, Riverside, had recently hired openly queer faculty members who organized seminars and conferences on gay and lesbian themes and incorporated queer scholarship into their courses. Some of the most intellectually exciting new faculty members were gay, and new queer-themed seminars, receptions, discussions, and performances appeared on campus each week, open to all. I attended many of the events on campus and made new friends. She was one of them.

Disturbed by this new attraction, I struggled to keep my feelings in check. I loved John and did not want to break up our marriage. And I was not sure what my feelings meant. Was it a momentary crush on this specific woman? Was I a lesbian? Bisexual? I had had passing attractions to both women and men. What was happening? I felt myself in the midst of academic erotic dances that I did not really understand. Hurt and angry about my emotional distraction, John confronted me one day about my "flirtatious behavior" with his colleague and demanded an explanation, wanting definite answers I could not give.

Despite our increasingly fraught relationship that spring, John and I decided to go to New York together for the six months of his sabbatical to try to figure out what was happening between us and to come to an understanding of what each of us wanted. We rented half of a loft in Soho owned by the artist Marcia Hafif, a friendly and intellectually adventurous monochrome painter who lived in the other half. Staying downtown on Mercer Street, with its cast-iron architecture in a neighborhood filled with artists we admired, we were in heaven and hell at the same time. Writing during the day, John and I went out at night with friends to as many theater and dance productions, art openings, readings, and performances as we could. We became "BAM slaves," as subscribers to the avant-garde Brooklyn Academy of Music called themselves. Yet much of the time we were miserable, each of us locked inside our own skins, feeling angry, betrayed, hurt, and alone.

On our return to Riverside in December 1991, John and I separated while we continued, with the help of a therapist, to see if we could salvage our relationship. I decided once more to try talking to my father. I did not initially tell him the reasons for our separation, only that we were struggling and going to couples therapy, which Dad, surprisingly, did not think productive. I must have been expressing anger at John, for my father said to me, "It's pointless to blame. You should start by asking yourselves how each of you may have been obstructing your relationship. Don't look at the other person; look at yourself."[9] I remember wondering with some trepidation how my eighty-four-year-old Freudian psychoanalyst father would respond to his recently married fifty-year-old feminist daughter coming out as a lesbian, which was how I was starting to think of myself. So one night late in the fall of 1992, I drove from Riverside to Los Angeles to have dinner with Dad at his Ocean Towers apartment. We sat at his dining table with the wine-colored linen tablecloth, light streaming from the dark green faux antique reading lamp he placed next to his plate so he could see, barely, what he was eating. It had been nearly two decades since homosexuality had been removed as a pathological diagnosis from the *Diagnostic and Statistical Manual of Mental Disorders* by the American Psychiatric Association and more than two decades since Stonewall had set off the gay and lesbian liberation movement of the late sixties and seventies. Freud himself had expressed a remarkably enlightened view for his time in a famous letter of 1935, written to the mother of a gay son. "Homosexuality is assuredly no advantage," he wrote to her, "but it is nothing to be ashamed of, no vice, no degradation; it cannot be classified as an illness; we consider it to be a variation of the sexual function, produced by a certain arrest of sexual development. Many highly respectable individuals of ancient and modern times have been homosexuals, several of the greatest men among them."[10]

I had heard Dad comment on same-sex attraction in casual conversation in ways that suggested his views were considerably outdated. Some of his early lectures described trying to encourage heterosexuality in his male patients and discourage or even forbid their same-sex "temptations," including those toward him.[11] He once said he had never seen a successful long-term same-sex relationship, though he agreed with me that he saw a distorted sample, since people in happy relationships would not seek therapy. More to the point, he believed sexual orientation was fundamentally defined by fantasy. What a person did in bed was far less important than what their fantasies were. So a man who slept with women all his life but who imagined, consciously or unconsciously, that he was with a man was, in Dad's view, gay. Similarly with women. He said that he thought great disparities between a person's sexual fantasies and their realities were problematic and that the person whose reality most closely approximates their fantasy—in the realm of sexuality, that is—is probably the best integrated.[12] But I knew that these views had been challenged as far back as the work of Alfred Kinsey in the 1950s and the studies of Masters and Johnson in the 1970s—certainly women's rape fantasies did not mean that women wanted to be raped, nor did fantasies of being whipped, tortured, beaten, or shamed, which were not uncommon, mean a person wanted to experience these acts in real life.

I still recall my father's sorrowful look when I told him I might be gay. "Alice," he said after a long pause, "you can't be angry at me for feeling sad that your life will be more difficult now." But that was all he said, and we did not discuss it further. He did not interrogate me about my feelings or my fantasies, as a therapist might have done. He did not analyze. He acted like a father, for which I was deeply grateful. After that night I noticed that he began to make positive comments about this or that person who was gay, whether living or dead. I had never heard him say such things before.

John and I eventually separated and ultimately decided to seek a divorce, though we remained on friendly terms. A year or two later when I showed up at Dad's apartment with a girlfriend, he tried to be welcoming, although his limited eyesight and hearing loss made making new acquaintances difficult. I remember thinking, *Freud's daughter Anna, also a psychoanalyst, shared her life with a woman, Dorothy Burlingham, for more than fifty years. I wonder if Dad knows.* But for some reason I never mentioned it to him, and we never discussed it.[13]

In March 1993, Nancy and the gene hunters announced that they had found "the most coveted treasure in molecular biology, the gene behind Huntington's disease." (The story appeared on the front page of the *New York Times* above the fold, in many other newspapers, and in the journal *Cell* under the collective authorship of the Huntington's Disease Collaborative Research Group.)[14] Their discovery made clear for the first time that the disease was caused by a gene that had expanded beyond a certain threshold; too many repetitions of a specific sequence within the gene (abbreviated as CAGCAGCAG, etc.) caused the protein (named huntingtin) encoded by the gene to have toxic effects in the brain. Although everyone has this gene, only some of us have the expanded version; too many CAGs mean that, sooner or later, you will develop the disease.

The New York Times

WEDNESDAY, MARCH 24, 1993

10-Year Search Leads Medicine To Elusive Gene

Scientists Find Cause of Huntington's Disease

By NATALIE ANGIER

After 10 backbreaking years in a research purgatory of false leads, failed experiments and long stretches of mordant despair, an international team of scientists says it has discovered the most coveted treasure in molecular biology, the gene behind Huntington's disease.

Now that they have the gene in hand, researchers say they can begin making headway in understanding the disorder, a neurodegenerative illness that usually strikes a person in the 30's or 40's, insidiously destroys body and sanity alike, and kills within 10 to 20 years.

Huntington's disease afflicts about 30,000 Americans, and as many as 150,000 are at risk of developing it. The best-known victim was the folk singer Woody Guthrie.

Holy Grail of Genetics

The first clues to the gene's location came in the early 1980's, at the dawn of the contemporary era of molecular genetics. But researchers soon ran into a succession of snags that transformed the search into an irresistible if irritating quest that seduced some of the biggest names in biology.

Of particular interest to scientists, the mutation that causes the disease is one they have lately seen in genes that cause other illnesses, a sort of molecular accordion effect in which a tiny segment of the gene is abnormally expanded and repeated over and over.

Key to the Puzzle

Researchers emphasized today that much work needed to be done before they could use the mutation as any sort of precise prognostic tool. Nor does the finding the gene mean that a treatment for the disease is imminent. But the discovery is essential to cracking the puzzle of Huntington's.

The finding will be reported Friday in the journal Cell, crediting as its author the Huntington's Disease Collaborative Research Group. This re-

President Boris N. Yeltsin of Russia held the hand of his wife, Naina, and fought back tears as his mot[...] was buried yesterday near Moscow. At the Kremlin, his power struggle with the Parliament continue[...]

Clinton Says He'd Consider Separating Gay Troo[...]

Remarks Are Contrast With Earlier Stand on the Military

By RICHARD L. BERKE
Special to The New York Times

WASHINGTON, March 23 — Facing resistance from the military over his pledge to allow homosexuals to serve in the armed forces, President Clinton said for the first time today that he would consider proposals to segregate troops by sexual orientation.

Mr. Clinton's comments at a White House news conference contrasted sharply with his public statements dur-

For Yeltsin's 'Coura[...] President Vows Pla[...] to Support Russia

By THOMAS L. FRIEDMAN
Special to The New York Times

WASHINGTON, March 23 — [Presi]dent Clinton brushed aside cri[...] today that he was too closely t[...] President Boris N. Yeltsin, sayin[...] Mr. Yeltsin's decision to circu[...] the Russian Parliament was "[...] priate" and that he would giv[...] Yeltsin "an aggressive and quit[...] cific plan" of aid at their summit[...] ing.

FIGURE 14.2 In the midst of the Bosnian war, power struggles in Russia, and President Bill Clinton's backing off his pledge to allow gays to serve openly in the military, the discovery of the Huntington's gene ranked as front-page news.

Like the marker discovery, this finding was also a landmark, revealing a whole class of neurological diseases whose origins lay in expanded CAG repeats. It immediately transformed Huntington's research. Suddenly it was possible for researchers to make animal and cell models and study how the gene worked at the cellular and molecular level. They could test drugs and other molecules in mice and sheep, fish and flies, as well as in human beings. They could investigate which ones might counteract the toxic protein or protect vulnerable neurons from its assaults. They could study mice and sheep well before they developed symptoms, track subtle changes in their brains in ways that were impossible with human beings. Not least, both predictive and diagnostic genetic testing suddenly became more accurate and much simpler, at least from a technical point of view. We even allowed ourselves to imagine that a treatment, and possibly a cure, might be on the horizon.

Dad was ecstatic and also relieved. He had feared that he might not live to witness this day. He was thrilled that the Huntington's story was attracting so much publicity and that Nancy was garnering increasing recognition. That year Columbia awarded her an endowed chair, making her the Higgins Professor of Neuropsychology. She also received the prestigious Albert Lasker Public Service Award. Soon she would be admitted to the National Academy of Medicine and to the American Academy of Arts and Sciences. She began to accumulate honorary doctorates, from the New York College of Medicine, the University of Michigan, Bard College, and Yale. I started to lose count of all her accolades.[15] These honors were usually presented at ceremonies that Dad and I both attended at the beginning. But after a while he could no longer come, so I alone accompanied Nancy and Herb. I loved being Nancy's sister at these events, where she would introduce me to all kinds of interesting people. Nancy always looked radiant, mingling with the likes of the former Supreme Court justice Sandra Day O'Connor and the architect Zaha Hadid just as she mingled with the beautiful, heartbreaking children of the families living with *el mal* in rickety tin shacks perched over the distant waters of Lake Maracaibo, far away but always close to her heart.

A slight disturbance in the air, a vague premonition, a perception so faint that I thought I must be imagining it, though I noted it in my diary at the time, on October 17, 1996. At the end of one of Nancy's visits to Los Angeles, I drove her to the airport for her flight back to New York, and I accompanied her to the departure gate (which could be done in the days before 9/11). We shared chicken salad in a box while we sat in the waiting area. Walking toward her after checking the boarding time, I noticed the awkward position of her legs angling sideways as she ate. The thought went through my mind, *Could Nancy have Huntington's?* I began thinking of other uncharacteristic aspects of her recent behavior, such as eating enormous quantities of sweets and not gaining weight and an atypical irritability, for she had berated the security guard at the gate when he asked her twice for her photo ID. I noticed no twitches or jerks, but I felt uneasy. I was scared.[16] I talked myself out of these concerns, and for a year or so my worries faded. But then another diary entry. I thought I had noticed Nancy moving around more than usual, not quite twitchy or jerky but wiggly, rocking back on her heels in a manner I had seen in some people with Huntington's.[17] On one of her visits to LA, she had expressed concern about her brain as we walked along the beach in Santa Monica, concern about spending too much money, doing too much. Within a few years some members of her Venezuela research team—world experts in Huntington's—began speculating aloud about Nancy having symptoms. There were phone calls and gossip. Nancy heard and became angry. She confronted the gossips. They were not her physicians. She was not their patient. They should not be making diagnoses outside the clinic. She wanted to shut them down.

I wondered what Dad noticed, given his dim eyesight. When Nancy and I watched television with him, sitting or lying with him on his bed, I would sometimes become conscious of Nancy's body moving next to me. She was wiggly. But I often felt wiggly too, I told myself. It was hard to know. Did Dad notice? He asked me once toward the end of his life, "Is Nancy OK?" I pretended he was asking about her financial situation and whether she had enough savings for the future. "Yes," I said. "She has an endowed chair as a professor at Columbia. She will be fine." He did not persist. But I knew what he meant. Dad said nothing more. He could not bear the thought. Neither could I. None of us considered the possibility of the genetic test to resolve the uncertainty. For all our knowledge of psychology, we turned to denial, that most primitive of defenses. We worried, we wondered, and then we denied. It simply could not be.

15
The Old Leaf

A round the time he turned seventy, my father began writing periodic farewell letters to Nancy and me. He wanted us to know that he preferred a party to any kind of ceremony. And that he had had "a very good ride throughout this life. There are some regrets but mostly I count my blessings." He told us that we were at the top of his list. I know he worried about how Nancy and I would manage when he died, with Huntington's never far from his mind. I do not know just when he wrote a proposal to create what he called the Great American Psychological Insurance Group, dedicated to giving its members the right to die with dignity and maximum choice. The group would function like a family and strive to prevent one another from being kept alive "even if we are vegetables" and to ensure that they would never become "receptacles for useless and agonizing procedures" by their doctors.[1]

Those words haunted me during my father's last stays at the UCLA hospital in Santa Monica in February and March 2007, when his lungs were giving out and I became preoccupied with the desire for him to have a beautiful death. But as things were going, that outcome did not seem likely. A rift was developing between my sister and me about how far to pursue treatment for our father's worsening condition. Dad himself seemed of two minds.

Four years earlier Dad had endured a botched surgery that had resulted in a colostomy, a risk about which we had not been informed before the operation and a heavy burden for a frail ninety-four-year-old man with limited vision and atrial fibrillation. Nancy and I wanted to sue the surgeon, but Dad was adamantly opposed. Former attorney that he was, he did not want to spend his last days embroiled in a lawsuit. We were fortunate that his savings and income, including Social Security, a pension, and continuing payments from a few patients, enabled us to hire round-the-clock home health aides. He could live in his rented apartment on San Vicente Boulevard, across the street from Palisades Park, and enjoy the "delicious experience" of listening to books on tape, especially the novels of Anthony Trollope, whom he considered a great psychologist. Such listening had become his greatest joy and often I would arrive at his apartment to find him in his recliner, eyes closed, hands clasped in his lap, listening peacefully to *Barchester Towers* or *The Eustace Diamonds*. He saw visitors, took walks, dictated letters, and watched—or rather, listened to—the news on television. He followed the research on Huntington's disease. He got to know his aides, appreciating their distinct talents and talking to one in particular about science. His mind remained sharp. But he who had always been fiercely independent now required full-time assistance, including with the daily, and for him, deeply humiliating, task of managing the colostomy bag. He endured it all with dignity and without complaint.

At the age of ninety-eight and a half, his lungs were filling up with fluid. The doctors diagnosed chronic obstructive pulmonary disease, COPD, damaged lungs "due to smoking," though he had quit forty years earlier. "Your father has a terminal disease," Dr. Jan Tillich, a UCLA cardiologist and Dad's primary physician, told us in early 2007. "He is going to die of this." The options were growing increasingly limited. The treatments to clear out his lungs were painful and only temporary in their effectiveness. Another pneumonia and more pulmonary obstruction lurked around every corner. Dad had always said that he did not want to end his life hooked up to a machine. A sudden death was preferable to a long and lingering one.

But now he was also saying that he wanted to live, with no mention of being tortured by doctors. Though sometimes he said it was fifty–fifty. "I'm a whipped puppy," he said on one especially difficult day. I wrote in my diary, "He basically cannot eat or drink anything without aspirating. Can't breathe on his own, can't eat or drink, can't swallow, so what kind of life is that at age 98-plus? Of course we could keep him alive with a tracheotomy and breathing tube and a feeding tube—but he'd basically be sedated—what pleasures would he have?" Even Frank Gehry, who visited Dad almost every day, said to us, "Maybe it's time." Dad himself recognized the dilemma. As he said one day in a voice to break your heart, "The thing that makes me better [food and drink] also makes me worse." From his hospital bed, he raised his arm into the air and, summoning a passage from his beloved Trollope, waved his hand slowly back and forth, lowering it gradually toward the floor. "You see this?" he said, describing an old leaf falling first this way, then that way. "I'd like to be that old leaf."[2]

A new CAT scan revealed that one of Dad's lungs was completely blocked by mucus plugs. He could not live long with just one lung, which was itself compromised. "He's going to be in trouble," said Dr. S., the pulmonologist. "He's going to fail." Dr. S. described an invasive procedure called a bronchoscopy that could clear out the blocked lung. Although the procedure took only about ten or fifteen minutes, Dad would have to be totally anesthetized and attached temporarily to a ventilator with a tube down his throat. Dr. S. opposed this choice on the grounds that Dad might die during the procedure. And if he did survive, he might never get off the ventilator and would be stuck, sedated, on a breathing tube, a machine, for the rest of his life, the thing he most feared and was determined to avoid.

We had the weekend to think it over and to discuss it with Dad, who was having difficulty focusing on the question. Nancy was convinced Dad wanted to live, and, as she said, "I have to respect that." She wanted to try everything possible for his recovery, including the bronchoscopy. I was not convinced Dad could recover and wanted to focus on his comfort. The phrase "let nature take its course" began to echo in my mind. Still, each day, each hour, was such a roller-coaster ride that even the doctors were surprised when Dad managed to pull back from the precipice just when they had been convinced he was about to go over. Bronchoscopy or no bronchoscopy. Neither option seemed the right one. I feared that no matter what we decided, our decision would be wrong, and our father would suffer even more.

On Monday morning the young resident, Dr. K., came into Dad's hospital room to explore his thinking about the procedure:

DAD: "Can you tell me if I have a fair possibility of living two more years? In your estimation, what percentage chance do I have of living two more years?"

DR. K., quietly: "A very low likelihood. A 25 percent chance of living two more years."

DAD: "What are the possibilities of lasting two more years if I have the procedure?"

DR. K.: "I don't think it would change. I don't think the procedure will add significantly to your life expectancy."

DAD: "So I'm at very high risk either way."

DR. K.: "Yes."

The room became very still. Dr. K. looked steadily at Dad, who gazed, unseeing, toward the windows. Nancy and I held our breaths.

Dad spoke quietly but firmly. "I accept that. I understand it. And therefore I do not want the procedure."

I did not tell my beautiful father that as we prepared to take him home, the pulmonologist said to us, "Enjoy the time you have left with him."

Two days later, Nancy and I were standing beside our father as he lay in his bed at home, his breathing becoming slower by the minute. It was five thirty in the afternoon. "Herring," Dad said faintly, according to Nancy, although I did not hear him. "Let's have herring and port at six o'clock." He liked having a glass of port each night before dinner, and we had made it a ritual whenever we ate with him. He who had taught us so much about living was teaching us now how to die. Maybe he would have a beautiful death after all.

Then time sped up and slowed down all at once. Kneeling on one side of the bed, Nancy wept and cradled Dad's head in her arms. "Dad, Daddy, come back, come back," she sobbed into his shoulder. "Daddy, don't die!" Kneeling on the other side, I put my cheek next to his and sobbed along with her. *Go to sleep you weary hobo.* I wanted to sing him the Woody Guthrie lullaby but I couldn't find my voice. "It's OK, Dad," I whispered into his ear. "We're going to be OK. It's all right to go." And then, and then, no more breaths. No more heartbeats. No more pulse. No more pain. No more Papa. "Every coming together is haunted by the inevitable separation," he had written to me nearly fifty years earlier when I was in the throes of youthful heartbreak. Now his words came flowing back into my mind. "Loving holds the painful promise that love or the loved one will eventually disappear. We just have the task of facing realities in that regard. Refusing to abdicate life, loving, and investing ourselves just because the end is always separation, or defeat, or death."

Dad wanted us to have a party after he died. His friend, the actor Carrie Fisher, who was not a patient, as she used to announce, offered her home in Bel Air. She had been visiting him and reading him drafts of her memoir, *Wishful Drinking*, and she had made him laugh. She had even come to see him in the hospital during his last weeks. I would have preferred a more intimate setting than Carrie Fisher's home, with its ambiance of celebrity. But her generous offer was hard to refuse. Many of Dad's friends, colleagues, patients, and former patients attended, as well as my friends and Nancy's and scientists associated with the foundation, some of whom had flown in from the East Coast, such as Anne Young, David Housman, and my childhood friend Emilie (Rahman) de Brigard. It all went by in a blur,

FIGURE 15.1 "Have a farewell party."

Photo by Elaine Attias, circa 1990s. Courtesy of Dan Attias.

though I remember Debbie Reynolds in a red dress, standing cheerfully at the entrance to the house (she lived next door) and as the guests departed, introducing herself to each one as "Carrie Fisher's mother," as if anyone did not know who she was, old Hollywood royalty. She was gracious and without pretense. The more meaningful farewell for me came two weeks later, when Frank and Berta Gehry took Nancy, me, and several others out on their boat from Marina del Rey into the Pacific late one afternoon. We all wept as we tossed yellow sunflowers—one of the few flowers bright enough that Dad could still dimly see before he died—along with his ashes, into the ocean. He would have loved it.

A front-page obituary in the *Los Angeles Times* proclaimed Dad "a visionary who led a genetic revolution." The *New York Times* credited him for the creation of the foundation's innovative science workshops and said that the research supported by the foundation had "changed everything in the world of genetic disease."[3] All the obituaries—and there were many—described Dad as a catalyst for bringing scientists from different fields together and for fostering new ways of thinking about research on disease. In the following days and weeks, letters and emails of condolence poured in, many from scientists who had attended foundation workshops but also from his former psychoanalytic colleagues, his students in Topeka, social workers, patients, artists, physicians, and Huntington's family members, as well as friends of Nancy and me. I knew that someday I would cherish them but not for a while. I needed to feel the pain of his absence, to absorb the reality that my father, who had seemed so permanent, so indestructible, as enduring and monumental as one of the faces carved into Mount Rushmore, was really gone from this earth.

In the months following Dad's death, I found myself ricocheting between sorrow and relief, relief that he was no longer suffering but also relief from anxiety and worry about him. I needed some distance. I needed to breathe. I wanted to immerse myself in the multitude of feelings that arose within me, the pain at losing his unconditional love, which I had never doubted no matter how fierce our arguments, and gratitude that he had lived long enough that our battles and misunderstandings had been, if not resolved, at least let go. As he grew older, his trust in me and respect for my writing, even when he disagreed with me, had lifted my self-estimation. He had given so much of himself to Nancy and to me, to Mom in her illness, and to the world. And now he was gone. I no longer had him as a bulwark between me and the universe, and I needed to experience this new raw way of being. Without children, Nancy and I were the last in our line, which made Dad's death seem even more of an ending. And although the idea of writing this book had been born so many decades earlier, I could not yet think of putting words to paper.

Finally, in 2014, I took the trip to Topeka where I began my quest to write about my father's life, especially his intellectual and professional life. As I searched the Menninger archives at the Kansas Historical Society and elsewhere, I also began poring through files at home, in my apartment, and in the foundation offices. Several letters turned up from his artist friend Jean-Jacques Lebel. I had always loved my father's story of stopping in Saint-Tropez in the summer of 1966, when Lebel had been staging Picasso's *Desire Caught by the Tail* on the beach. So one day after the close of a Huntington's meeting in The Hague in the summer of 2015, I boarded a train for Paris to visit Lebel. I wanted to talk to him about his friendship with Dad.

"We hit it off immediately," Lebel told me. "He was not like these French analysts who always try to corner things and define things into small boxes. He had
a much more free-flowing perception of what human thinking was all about."
I asked him what he thought about my father coming to see the Picasso play in
the sixties. "Look, there aren't many LA psychiatrists who would do that, right?"
Lebel said. He had found Dad "extremely refreshing." Like most of his generation, he'd been through psychoanalysis, but except for Félix Guattari (a distinguished French analyst), he'd never encountered that freshness of approach, "that
openness to the inner workings of poetry and the inner workings of the art-pro-
ducing machinery of the mind. You know, the thing that John Cage talks about
so brilliantly when he says, 'It's the process that counts; it's not the result.'" Lebel
believed my father understood that idea instinctively.

Then he told me a story about himself, my father, and a famous painting
called *The Tempest* by the sixteenth-century Venetian, Giorgione, which hangs
in the Accademia Gallery in Venice. Lebel was in love with that enigmatic painting. In it a nearly naked woman sits near a riverbank, with a town in the distance, nursing her infant and looking at the viewer, while opposite her stands a
youth, possibly a soldier, with a lance, looking at her. Lightning flashes across a
sky dark with storm clouds. "I've been studying that painting all my fucking life,
and I'm still wondering what it's about," Lebel said. He told me he remembered
sitting with my father in front of the painting, the two of them meditating on it
because it was such a powerful statement "of subconscious feelings, desires, and
vistas of humanity, [of] what really goes on in people's lives. It's so beautiful.
It's pure Renaissance." It was only a small painting, he said, but it could take a
lifetime to understand. "It is one of those things that haunts you forever. Like
Shakespeare, like Rimbaud, and I found it so refreshing that I could share this
with an American person, from California, Hollywood, and all that stuff. For us
French people, Hollywood is like some kind of circus. And here's this guy who
knew where to look."[4]

Frank Gehry told me, "He was as much an artist as any of us."[5]

Epilogue

April 6, 2020. I am sitting at my desk in Santa Monica sheltering in place and social distancing, as they say, trying to follow the rules for flattening the curve of the coronavirus pandemic that is devastating the world. Nancy has been diagnosed with Huntington's disease, still living her life with courage and resolve to wrest from her days and nights every possible moment of joy. Our father has been dead for thirteen years. He had wanted his ashes scattered in his beloved rose garden overlooking the Pacific, in Palisades Park here in Santa Monica, where he had walked from his apartment so many times. He told Nancy and me that his only concern about having his ashes among the roses was that it might create a sense of obligation to visit from time to time. "Forget that," he wrote to us. "The world is filled with too many obligations. I don't wish to add to that scenario." But when we were confronted with a difficult problem, maybe we could find the rose garden an inspirational location at which to think it through.[1]

Although his ashes are in the ocean—which was also quite acceptable, he told us—the rose garden remains my sanctuary and thinking place. I go there from time to time to recall his many faces and voices, his fierce love and his formidable anger, his smiles and his scowls, his great sense of responsibility but also his determination to live life on his own terms; his equal determination that we, his daughters, Nancy and I, be free to pursue our dreams. My dream of writing books, even a book about him.

I would like to visit the rose garden today, but it is closed, as is Palisades Park and all the beaches and the bike path below, to keep us from infecting each other with COVID-19 should we walk close together or sit side by side, as my father and I did over so many years. One day this pandemic will be over, but nothing will be the same, except for the memories and images of my father that live in the pages of this book that I began so long ago and that now has come to pass.

Acknowledgments

My father's willingness to lend his life to his critical daughter's scrutiny, starting in 1969, was the essential foundation for this book, a gift for which I am endlessly grateful. Although my plan to write his biography lapsed, this memoir, begun more than forty years later, could not have existed without his generosity and support. A superb editor, Jackie Wehmueller, believed in the project when I presented it to her in 2018. It was she who helped structure and shape the narrative, teaching me about writing and thinking and the wondrous value of a great editor along the way. Working with her was a joy and a privilege. I am also immensely grateful to my cousin Eric Wexler for having the wisdom and foresight to preserve his father's journals and my father's letters to my uncle bundled with them and for his generosity in making both available to me. Without them, this book would have been very different. I also thank my cousin, the writer Jane Stern, for sharing priceless family photographs and trading family lore. Dan Attias generously returned the letters Dad had sent to his mother, Elaine. He also shared photographs taken by Elaine during their time together and allowed me to reproduce some of them.

Over many years my brilliant writing group sisters—Ellen Krout-Hasegawa, Kathleen McHugh, and Harryette Mullen—have indelibly shaped my thinking with their empathic questions, imaginative suggestions, and generous companionship. Their company feeds and sustains me. My historians' "Supergroup"—Emily Abel, Carla Bittel, Charlotte Borst, Janet Brodie, Sharla Fett, and Janet Golden—has also provided invaluable insights and critiques that have guided this book and helped sustain a long-haul journey through the pandemic. At an early stage of the writing, Sara Melzer invited me into her legendary weekend writing retreats with the writing coach William Waters, whose guidance informed the initial outlines of the book. At a late date, Ellen Krout-Hasegawa reviewed the entire manuscript, twice, offering astute feedback and suggestions.

My father's friends, colleagues, students, and former patients, and their spouses—and some of their children—generously shared memories from distinct stages of his life. I am especially grateful to Gerald and Jan Aronson, Bud Cort, Jodie Evans, Frank and Berta Gehry, Alfred Goldberg, Arthur Golding and Dewanna Covey, Hildi and Joan Greenson, Robert Holt, Maimon Leavitt, Jean-Jacques Lebel, Ed Moses, Herbert Schlesinger, Julian Schlossberg, Roberta (Babs) Thompson, Lisa Weinstein, Miriam Wosk, and Robert Wallerstein, whom I was lucky enough to correspond with and interview in his Belvedere, California, home just months before he died in December 2014. Frank Gehry has shared his clear-eyed and penetrating views of my father over many years in conversations that have been one of the pleasures of this project.

For sharing their knowledge of Theodor Reik, I thank Dany Nobus of Brunel University and Martin Shulman of the National Psychological Association for Psychoanalysis, who guided me through the Douglas Maxwell Library. David Carlson at the Western New England Institute for Psychoanalysis offered useful information about William Pious, my father's closest Topeka friend. Lawrence Friedman, the author of the definitive work on Menninger, introduced me to the vast and mysterious Menninger Foundation Archives at the Kansas Historical Society. Moshe Sluhovsky, a historian at the Hebrew University of Jerusalem, awakened me to the work of Dagmar Herzog, whose elegant work *Cold War Freud: Psychoanalysis in an Age of Catastrophes*, along with Eli Zaretsky's *Secrets of the Soul: A Social and Cultural History of Psychoanalysis*, and *Political Freud: A History*, provided essential context.

I am grateful for the assistance of the librarians, archivists, and researchers who tracked down unsuspected letters of my father and references to him in a number of archives and libraries. Many thanks to Laura Sklansky, an independent researcher who scoured the Chicago History Museum and Chicago Public Library for correspondence to and from Eppie Lederer (Ann Landers); Lizette Royer Barton at the Drs. Nicholas and Dorothy Cummings Center for the History of Psychology at the University of Akron, in Akron, Ohio; Nellie Thompson at the Abraham A. Brill Library of the New York Psychoanalytic Society and Institute; Michael Gales and Mal Hoffs for facilitating access to the LAPSI archives many years ago; Chamisa Redmond and Margaret Kieckhefer at the Manuscript Division of the Library of Congress; Russell Johnson of the Louise M. Darling Biomedical Library History and Special Collections for the Sciences at UCLA; and the archivists at the Menninger Foundation Archives at the Kansas Historical Society in Topeka. I also appreciate the invaluable work of Sam Markham, Jean Moylan,

Kate Kirwan, and Suzana Chilaka, archivists from the Winthrop Group who inventoried Nancy Wexler's enormous archive and identified important sources. Many thanks also to Marianna Cook who allowed me to reproduce her photograph of Nancy, Dad and me; to Margo Howard for permission to reproduce one of her mother's letters to my father; and to the family of Ken Price for permission to reproduce his iconic portrait of John Altoon with Dad.

Writing is lonely and was especially so during the height of the coronavirus pandemic of 2020 to 2022, when socializing was severely curtailed. I owe a huge debt of gratitude to friends who are writers, readers, scholars, and artists of many kinds for conversations that took place outdoors. If there was any personal upside to the pandemic, it was hiking the deep canyons and high ridges of the Santa Monica mountains and walking all over Los Angeles with friends while talking about writing, the world, and ourselves. A special thanks to hiker-historians extraordinaire Alice Echols, Soraya de Chadarevian, Robert Rosenstone, Marla Stone, and Devra Weber for insights into the social dimensions of science, Los Angeles history and politics, and the complexities of melding memoir with historical narrative. Conversations with the literary wing of the Will Rogers hikers Doris Baizley, Katherine King, and Felicity Nussbaum were the best possible incentive for getting out of bed in the morning; I am grateful for endlessly interesting and enlightening discussions of writing for the stage, the page, the street, and screens of all kinds.

I also thank fellow walkers and hikers Karen Brodkin, Michael Cahn, Michael Cowan, Scott Johnson, Nahid Massoud, and Ella Taylor for lifegiving companionship and conversation through the seemingly endless days of the pandemic. A special appreciation to Ann Goldberg for sharing the pleasures and perils of our psychoanalyst fathers, to Yong Soon Min for joyful hikes to the turtles atop "Mt. Everest" at Ernest E. Debs Regional Park, to Kate Flint for untying the mother knot, to David and Alexander Callahan and Wendy Paris for superlative cocktails and conversation, to Joe Boone for heavenly dessert deliveries, and to Ian Song whose phone calls and humorous texts always brightened difficult days. Herb Pardes generously put up with visits amidst Covid and other forms of chaos.

Zoom also came to the rescue during the pandemic, allowing essential and entertaining weekly conversations with Nancy Wexler, Jenny Morton, Sir Michael Rawlins, and Anne Young, who also answered every panicked phone call with her trademark wisdom, humor, and empathy. Zoom dinners shared with Muriel McClendon and Sara Melzer introduced me to new culinary delights while offering great company in lonely times. When it was safer to travel to New York, I feasted on conversations with the authors of classic memoirs Vivian Gornick and Penny Wolfson, whose art of conversation equals that of their elegant writing.

Dan Rubey, a former fellow BAM slave, and Karen Sughrue shared discoveries about art and politics from their imaginative explorations in New York, and drinks with Irene Zola at Le Monde never failed to cheer. Conversations over the years with Rosemarie Garland-Thomson, many on walks around the UCLA campus before the pandemic, taught me to question representations of impairment and to think differently about disabilities of all kinds. She challenges me and so many others with her radical critique of bodies and her reimagining of the possibilities for a good life. I also thank longtime friends Stephanie Kay and Carole Fabricant for their encouragement and sense of humor in dark days, and Christina Benson, who got me back on track when I went off the rails at painful moments. The intellectual daring and playful imagination of Emilie Hermant and of Dingdingdong have a special place in my heart. Thanks also to my essential extended Hereditary Disease Foundation family at 168th Street, Julie Porter, Meghan Donaldson, and Kenneth Martin.

I had the privilege of a dream team at Columbia University Press, what every author wishes for. Miranda Martin skillfully laid the groundwork for what seemed to me an overwhelming process, while her assistant Brian J. Smith had a prompt solution to any and all dilemmas. Michael Haskell directed me to a wonderful production team at KnowledgeWorks Global Ltd., where Kalie Koscielak managed a complex operation with aplomb and patience. My terrific copy editor, Kara Cowan, not only corrected grammar and brought clarity to obscurity but caught embarrassing factual errors and inappropriate language. The art department brought prodigious skills to all aspects of the images and design. I owe a special debt to the anonymous reviewers of this manuscript, for their suggestions and corrections, and to Mari-Jo Buhle, Alice Kessler-Harris, Ellen Herman, and Alexandra Stern for their generosity of judgment.

While she did not always share my view of our father and even objected at times to my writing, my sister, Nancy, has always been my closest confidante and most beloved comrade, though we have lived much of our lives three thousand miles apart. Her genius transformed the contemporary world of genetics and neuroscience. She also comforted thousands of people living with Huntington's, mentored hundreds of young scientists, and indelibly shaped our father's life and mine. Her vision, love, and courage in the face of her own diagnosis of Huntington's inspire and motivate me every day.

APPENDIX

Sayings of Milton Wexler

S ome of my father's friends and former patients told me that, at times of trouble, they would ask themselves, what would Milton say? And so I wanted to end this book with passages from his letters and lectures that are seared into my brain for their wisdom, their clarity, their Milton-ness.

FROM LETTERS

ON THE LOSS OF A CHILD

There is no tragedy to equal this. I would not dare say comforting words. A loss of a child is unimaginable. It is unacceptable. It's an indelible stain on one's joy in life. Even my mind cannot embrace it, and I didn't know your son.

(To a grieving father, July 1, 1998)

ON SELF-DEFEAT OR DEFEAT AT THE HANDS OF OTHERS

Whatever you do, do not defeat yourself by abdication, by despair, by self-immolation in a storm of self-flagellation. Just go about your business. Just do your job as best you can and try to keep the garbage as much out of your mind as possible. If you are to be defeated, be defeated by your enemies, not by yourself. No amount of self-accusation, confession, or tragic posturing will ever satisfy the jackals who are going after your hide. Let them go about their sinister behavior. Let them play out their game of the scarlet letter. Let them have their Grand Inquisitor. You take care of the affairs of state.

(To President Bill Clinton, December 19, 1998)

ON HOLIDAYS

Basically, I hate holidays. They have been taken over by corporate enterprise and the advertising mafia. But it is nice to get a card or a letter or a picture, or any other kind of hello. It gives coherence to life. It connects one with one's past. It is good for the soul.

(To an acquaintance, January 5, 1996)

ON PSYCHOTHERAPY

I often say that I am not a psychotherapist. I would describe myself as, perhaps, a teacher or a philosopher or a fellow sufferer. I tell patients that it is up to them to get themselves better, improve their lives, or make relationships more workable. I can react more honestly to them than most of their friends or colleagues. It's up to them to make use of that. I can provide them with the fruits of my experience and my observations, but I cannot guarantee relevance beyond that which can be found in a very good novel by Trollope or Dostoyevsky.

(To Richard Farson, July 16, 1996)

ON IDENTITY

I don't think that one ever really solves that. We all have "relative" identities. Relative to the time, place, person, and conditions of the moment. I think you have a well-crystallized identity in some areas, and in some you are still searching. So am I. So what? Isn't that part of the fun of living? The kind of thing that refuses to have hardening of the arteries or of the identity, that can say, "I can change!"

(To "Dear Pooch," n.d.)

ON LIFE AS STRUGGLE

I'm glad you have some good friends to fall back on. That is as much as we can demand in life, which seems always to be a series of disasters with interludes of surcease and pleasure. I'm not such a pessimist as I sound. There is a way of looking at all this which is far from pessimistic. It is rather in the direction of a gallant struggle against great odds, and if you come out on top or near enough the sun to get a good psychic tan it is more than enough. I've long ago given up thinking in terms of ultimate victories, ultimate accomplishments, ultimate relationships and learned rather to expect and even enjoy the process of overcoming difficulties and leading a rather day-to-day life of matching my resources against what life has to offer and signing off each day with some real satisfaction if I've given it the best I have. Not much in the way of profound wisdom or lightning-bolt insight. But for the moment it is the way I feel rather than think.

(To Alice Wexler, 1967)

ON LONELINESS AND ALONENESS

You speak of the problem of feeling alone. I think most of us feel pretty much alone. It is damnably hard to be understood by another. . . . All one can do is develop the kinds of tastes and interests that one believes in or enjoys . . . and then see if you can uncover another person who swings to the same tune, or develop some-one who learns to do it. But at most that is a way of cutting out the general state of aloneness that always persists to some extent. And even that state has its own compensations because it can be seen in some lights as the heroic struggle we all go through to overcome the barriers erected by our skins and minds.

(To "Dear Pooch," n.d.)

ON REASON

I think you will find that reason persuades very few people. We are all too primitive for that. We are too guided by the unconscious, whether as individuals or as groups (nations). I also would like to approach people on the basis of reason, rationality, logic. I'm best at that. But I know that even in the most educated and sophisticated groups (see LA Psychoanalytic Society), reason and logic are highly secondary to primitive drives, power motives, suspicion of the stranger, competitive anxieties, etc. The most persuasive communication is example: how you behave, think, feel.

(To "Dearest Pierced Ears," November 1964)

ON SUCCESS AND ATTAINMENT

Attainment is not necessarily related to the society around. In fact it can be an attainment geared to oppose the society. Or it can be an attainment that has only to do with the individual and spells out his significance for himself.

(To Alice Wexler, July 20, 1968)

ON RISK

There is always risk in living. And if some increment of risk is added to your life [Huntington's], all one can do is learn to live with it. To collapse into the mothering arms of a therapist merely intensifies the sense of risk; it does not soothe. I know that for sure.

(To Alice Wexler, July 12, 1971)

ON BEING A FATHER

[It] seems to me my involvement in CCHD, unhappy though the causes, brings us all closer in some ways. If I'm super busy, and you are both away, does that make me less a father? Or is it that you're afraid that I'll pop off from overwork? Never mind. It's not important. I'll probably be the only kind of father I know how to be under any circumstances.

(To Alice Wexler, November 1971)

ON RELATIONSHIPS

I think one must recognize something about the human condition, namely, that a truly creative, gratifying, synchronized, and satisfying relationship or relationships are hard to come by.

(To Alice Wexler, November 20, 1965)

ON LOVE

On the whole I am inclined to believe that it is better, especially when one is reasonably young, to shoot for the moon. That is, to try to make a deep and abiding relationship on the basis of real love and real commitment. For me, and I guess for most people, that is the very best. It is the most rewarding, potentially, even though the most difficult; it is also the most painful if it doesn't work out. But big gains are obtained only by big risks.

(To "Dearest Pooch," n.d.)

ON GIVING A SERIOUS DIAGNOSIS TO A PATIENT

What I've learned is that people respond with fright and denial when doctors and scientists are tentative and cautious, overly tender and concerned. When they are straightforward, blunt, realistic, and detailed in their explanations, most patients and families respond with marvelous courage. What is true of patients and their families would be true of the public at large, I'm sure.

(To Harry Levinson, December 15, 1970)

ON COMMUNICATION ACROSS BARRIERS

While we are mainly imprisoned within our own skins and can only rarely in life reach across barriers to really share and be with another person or group, there is something really great and worthwhile in making the persistent effort to reach across that boundary. I don't have to remind you that even the people closest to each other, who love each other the most, will experience distance and hurt and suspicion and anger. Communication is an art, love is an art, understanding is an art, and it takes all these artistic talents to bridge the gap of suspicion and doubt and selfish, individualistic motives in order to establish something like mutuality.

(To "Dearest Pierced Ears," November 1964)

ON SELF-ACCEPTANCE

The big battle is always centered around accepting oneself. Self-criticism is always more painful in the long run than the criticism of others, in fact is what makes criticism from the outside seem so dangerous.

(To Alice Wexler, 1965)

If we would only learn to maximize what is unique in us we'd all be less carbon copies, less programmed, less competitive.

(To "Dearest Dolls," January 14, 1969)

ON DEPRESSION

You're right about the struggle with depressions that most of us go through all our lives in one way or another. It's another variant on the struggle to find the right and best and most loving object, the aspects of mourning for the ideal of childhood. We never do find that perfect lover, the perfect understanding, that perfect intimacy, but some of us are never reconciled and so go on with our periodic depressions, hoping. And perhaps, like your present life as expressed in your letter, learning that the best way is to keep on pressing for the best in our present situation, our present relationships. And not being afraid that we are settling for less (less than the perfection of infancy?) or that we won't make it (make what? the post of favorite child?).

(To "Dearest Ali," 1969)

ON LIVING

Being your own person has a deep meaning. Search for that. I think you will find it and get some joy from it.

(To "Dearest Ali," January 1971)

Revolutions against anything are always half-assed. It's what you are for that counts.

(To Alice Wexler, August 1964)

ON TEACHING

The right questions are more important than the right answers. Enthusiasm is more important than brilliance. Curiosity is more important than profundity. One must have trust that the student can also get things for himself, find answers, attain depth. As in therapy, it is sometimes more important to stay out of the way of learning than to feed it.

(To "Dearest Ali," October 22, 1967)

FROM LECTURES AND TALKS

"NOTES ON THE PSYCHOTHERAPY OF SCHIZOPHRENIA," JANUARY 28, 1953

I know many patients with very bizarre distortions of thought and feeling who have made extraordinarily successful adaptations to an outside living situation.

Frequently they give explicit indications that the trust and confidence placed in their capacity to adjust, even to conceal the psychotic manifestations where necessary, represented the sole support needed for success.

In this acute phase, then, I have been more impressed with the importance of indicating to the patient that he is understood.

"THE PSYCHOTHERAPY OF SCHIZOPHRENIA," LECTURES AT THE LOS ANGELES PSYCHOANALYTIC SOCIETY AND INSTITUTE, OCTOBER 13 TO OCTOBER 20, 1958

I believe the risk is greatest when we attempt to preserve our science and scientific attitude rather than to preserve the patient.

As a general principle: Begin with the patient each session where he is and not where you would like him to be.

In all humility [I] may say that special skill involves permitting self-cure; just don't interfere with restitutions.

Confidence and consistency in practice may be as important as rightness of theory. A mechanical view of physics was incomplete and even incorrect but served many practical purposes.

[There is] no better way to establish contact than to put into words some subjective feeling of theirs which they are afraid to mention or haven't yet put into words.

"THE PSYCHOTHERAPY OF SCHIZOPHRENIA," LECTURE AT THE UNIVERSITY OF SOUTHERN CALIFORNIA MEDICAL SCHOOL, 1964

I am inclined to give very great significance to Eissler's comments here that affect is the real element and that communication and identification and understanding (even of the most primary process language) come by way of affective communication.

"A TRIBUTE TO ANNA FREUD ON HER SEVENTIETH BIRTHDAY," 1965

The most meaningful tribute we pay to another person lies most certainly in the effort we expend to understand them.

"PSYCHOANALYTIC THEORIES OF SCHIZOPHRENIA," LECTURE AT THE CITY OF HOPE NATIONAL MEDICAL CENTER, 1973

Naturally poets and philosophers got there first. Witness Nietzsche in advance of Freud on dreams and conflict. Goethe: "If the sun be not within us, how then shall we see the sun?"

Published Papers and Reviews by Milton Wexler

1947. "Psychiatric Screening of Naval Personnel," in *New Methods of Applied Psychology: Proceedings, Maryland Conference on Military Psychology*, ed. G.A. Kelly (College Park, MD: University of Maryland Press).

1947. "Measures of Personnel Adjustment," in *Personnel Research and Test Development in the Bureau of Naval Personnel*, ed. Dewey B. Stuit (Princeton, NJ: Princeton University Press), 126–76.

1948. "Psychotherapy and Counseling: Summary of Discussion," ed. R. Nevitt Sanford, *Journal of Consulting Psychology* 12 no. 2: 88–91.

1951. "The Structural Problem in Schizophrenia: Therapeutic Implications," *International Journal of Psychoanalysis* 32, no. 3: 157–66.

1951. "The Structural Problem in Schizophrenia: The Role of the Internal Object," *Bulletin of the Menninger Clinic* 15, no. 6: 221–34.

1952. "The Structural Problem in Schizophrenia: The Role of the Internal Object," in *Psychotherapy with Schizophrenics*, ed. Eugene B. Brody and Fredrick C. Redlich (New York: International Universities Press), 179–201.

1953. "Mental Hygiene and the Muscular Dystrophy Patient," in *Proceedings of the First and Second Medical Conferences of the Muscular Dystrophy Associations of America Held in New York, NY, April 14–15, 1951 and May 17–18, 1952*, ed. Ade T. Milhorat (New York: Muscular Dystrophy Associations of America), 58–67.

1953. "Psychological Distance as a Factor in the Treatment of a Schizophrenic Patient," in *Explorations in Psychoanalysis: Essays in Honor of Theodor Reik, on the Occasion of His Sixty-Fifth Birthday, May 12, 1953*, ed. Robert Lindner (New York: Julian), 157–172.

1960. "Hypotheses Concerning Ego Deficiency in Schizophrenia," in *The Out-Patient Treatment of Schizophrenia*, ed. Sam C. Scher and Howard R. Davis. (New York: Grune & Stratton), 33–43.

1962. "A Temple for Rosen?" *Contemporary Psychology* 17, no. 5: 171–73.

1965. "Working Through in the Therapy of Schizophrenia," *International Journal of Psychoanalysis* 46, no. 3: 279–86.

1966. "Comment on Dr Bychowski's Paper," *International Journal of Psychoanalysis* 47, nos. 2–3: 198–202.

1967. "The Scientific and Professional Aims of Psychology," in *American Psychologist* 22, no. 1: 49–76 (participant in panel discussion).

1968. "One Path to Recovery Lies in the Development of Stable Object Relationships with Someone in the Outside World," in *Treatment of Schizophrenia: A Comparative Study of Five Treatment Methods*, ed. Phillip R. A. May (New York: Science House), 278–84.

1969. "The Non-transference Relationship in the Psychoanalytic Situation," *International Journal of Psychoanalysis* 50, no. 1: 27–39 (with Ralph R. Greenson).

1970. "Discussion of 'The Non-transference Relationship in the Psychoanalytic Situation,'" *International Journal of Psychoanalysis* 51, no. 2: 144–50.

1971. "Psychoanalysis as a Basis for Psychotherapeutic Training," in *New Horizon for Psychotherapy: Autonomy as a Profession*, ed. Robert R. Holt (New York: International Universities Press), 184–193.

1971. "Schizophrenia: Conflict and Deficiency" *Psychoanalytic Quarterly* 11, no. 1: 83–99.

2002. *A Look Through the Rearview Mirror* (New York: Xlibris).

List of Abbreviations

The following abbreviations appear in the notes:

auto. tape	Milton Wexler's autobiographical tape, author's collection
DRP	David Rapaport Papers, part 1, container 1, Folder 32; Sigmund Freud Collection, Manuscript Division, Library of Congress, Washington DC
HDF	Hereditary Disease Foundation
LAPSI	Los Angeles Psychoanalytic Society and Institute
MFA	Menninger Foundation Archives, Kansas Historical Society, Topeka, Kansas
Research Dept.	Records of the Research Department, Menninger Foundation Archives, Kansas Historical Society, Topeka, Kansas
RVM	Milton Wexler, *A Look Through the Rearview Mirror* (self-pub., Xlibris, 2002)
SP	Schizophrenia Project, Records of the Research Department, Menninger Foundation Archives, Kansas Historical Society, Topeka, Kansas
"WW"	Milton Wexler, "Winning Ways" (unpublished manuscript, 1998), author's collection

Notes

1. ON THE ROAD TO TOPEKA

1. See Lawrence J. Friedman, *Menninger: The Family and the Clinic* (New York: Knopf, 1990); Edith Beck to Karl Menninger, Nov. 22, 1951, MFA; Harold Maine, "We Can Save the Mentally Ill," *Saturday Evening Post*, Sep. 15, 1947; see also Irwin Rosen, "Psychoanalysis at the Menninger Foundation: Beginnings, Growth, and National Influence," Oral History Workshop #14, San Juan, Puerto Rico, May 7, 1981; *Psychohistory Review* 19, no. 1 (1990), special issue devoted to Menninger.
2. Information about Nedda from Milton Wexler, monthly reports, November 1948 to June 12, 1951, SP, Research Dept., MFA; also U.S. federal census, army enlistment records, https://www.ancestry.com.
3. Robert P. Knight, introduction to *Psychotherapy with Schizophrenics: A Symposium*, ed. Eugene B. Brody and Frederich C. Redlich (New York: International Universities Press, 1952), 11.
4. Joel T. Braslow, "Psychosis Without Meaning: Creating Modern Clinical Psychiatry, 1950 to 1980," *Cultural Medical Psychiatry* (2021): 451–53, doi.org/10.1007/s11013-021-09744-3; Braslow, *Mental Ills and Bodily Cures: Psychiatric Treatment in the First Half of the Twentieth Century* (Berkeley: University of California Press, 1997); Jonathan M. Metzl, *The Protest Psychosis: How Schizophrenia Became a Black Disease* (Boston: Beacon, 2009), 211; Jack D. Pressman, *Last Resort: Psychosurgery and the Limits of Medicine* (Cambridge: Cambridge University Press, 2002). On Karl Menninger's therapeutic optimism, see Karl Menninger, "The Diagnosis and Treatment of Schizophrenia," *Bulletin of the Menninger Clinic* 12, no. 3 (May 1948): 96–106; Friedman, *Menninger*, 46.
5. John N. Rosen, "The Treatment of Schizophrenic Psychosis by Direct Analytic Therapy," *Psychiatric Quarterly* 21 (Jan. 1947): 117–19; reprinted with discussions in John N. Rosen, *Direct Analysis: Selected Papers* (New York: Grune & Stratton, 1953), 44–75; Milton Wexler to Sibylle Escalona, Sep. 20, 1948, SP, Research Dept., MFA.
6. The initial observer was David Rubinfine, followed by George Harrington when Rubinfine left Topeka.
7. *Object* in this context refers to a person; an object in Freudian theory may be a person, a part of a person, or a symbol of such that is the target of a person's instinctual impulse such as desire or aggression. An object may be an entity in the outer world or an internal representation.
8. Milton Wexler to Sibylle Escalona, Sep. 20, 1948, SP, Research Dept., MFA.
9. Monthly reports, Nov. 8 to Dec. 1, 1948, Dec. 1948, Feb. 1949, SP, Research Dept., MFA.
10. Monthly report, Mar. 1949, SP, Research Dept., MFA; Milton Wexler, "The Structural Problem in Schizophrenia: Therapeutic Implications," (unpublished manuscript, Jun. 1950), author's collection.

11.   Monthly reports, Jul., Sep., Oct. 1949, SP, Research Dept., MFA.

12.   Monthly reports, Mar., May 1950, SP, Research Dept., MFA. Mary Leader was the social worker.

13.   Milton Wexler, "The Structural Problem in Schizophrenia."

14.   Pressman, *Last Resort*; Gail A. Hornstein, *To Redeem One Person Is to Redeem the World: The Life of Frieda Fromm-Reichmann* (New York: Other, 2000), 148–49.

15.   Monthly reports, May, Jul., Aug., Sep. 1950, SP, Research Dept., MFA.

16.   Monthly reports, Nov. 1950, SP, Research Dept., MFA.

17.   Sibylle Escalona, test report, Nov. 1950, SP, Research Dept., MFA.

18.   Monthly reports, Sep., Oct., Nov. 1950, SP, Research Dept., MFA.

19.   See Eugene B. Brody and Frederick C. Redlich, eds., *Psychotherapy with Schizophrenics: A Symposium* (New York: International Universities Press, 1952).

20.   John Rosen was eager to have his approach "scientifically evaluated," as he told Sibylle Escalona during his visit. He may have helped initiate the discussions that eventually led to the formulation of the project; in its earliest iterations he was designated chief therapist, although my father soon took over that role.

21.   See Robert P. Knight, introduction to *Psychotherapy with Schizophrenics*. The Menninger archival material suggests that the idea for the project may have emerged during John Rosen's visit to Menninger in January 1948, when he broached it with Sibylle Escalona and others. It was discussed more formally in April 1948 at a meeting of the Research Committee of the Group for the Advancement of Psychiatry, in Asbury Park, New Jersey. At this meeting Rosen again discussed his work, after which the committee—and possibly Sibylle Escalona—drew up a plan for studying his methodology using the facilities and resources available at Menninger. Sibylle Escalona's proposal of April 13, 1948, centered on John Rosen as the chief therapist with "a senior research psychologist, to be designated, who will be responsible for observing and recording those aspects of the treatment which are not obtainable through sound recording." One week later, on April 22, her letter to my father referred to "our 'projected project'" and indicated that he was to be part of it, along with Rosen if Rosen was willing; Sibylle Escalona to William C. Menninger, Jan. 26, 1948; Sibylle Escalona to Milton Wexler, Apr. 22, 1948, SP, Research Dept., MFA. When Rosen was unable to participate or decided against participating, my father drew up his own proposal for "the treatment of a selected few schizophrenics with direct analytic therapy" that did not depend upon Rosen's involvement. The proposal went through many iterations and involved extensive correspondence and many memos. See especially Sibylle Escalona, "Proposed Research Project," Apr. 13, 1948; Sibylle Escalona and Milton Wexler, "Preliminary Information Regarding Current Plans to Study Dr. Rosen's Method of Psychotherapy with Schizophrenic Patients," Apr. 13, 1948; Sibylle Escalona and Milton Wexler, "Preliminary Statement of Research Plans on the Treatment of Schizophrenic Psychosis by Direct Analytic Therapy," May 6, 1948; Milton Wexler to "Manager," Jun. 3, 1948; Milton Wexler, "The Therapy of Schizophrenia" in Peter D. Fleming, *Information Bulletin*, Winter Veterans Administration Hospital Research Board, Jul. 1, 1948; Milton Wexler to Sibylle Escalona, Oct. 5, 1948; Sibylle Escalona to Milton Wexler, Oct. 8, 1948; Milton Wexler to Sibylle Escalona, Oct. 10, 1948; Milton Wexler, "An Investigation Into the Treatment of Schizophrenic Psychosis and the Nature of Schizophrenia by a Direct Analytic Method," May 25, 1950. For fundraising appeals, see William C. Menninger to George S. Stevenson, Nov. 5, 1948, Nov. 6, 1948; William C. Menninger, June 6, 1949; Sibylle Escalona to Leo H. Bartemeier, Sep. 20, 1949; all from SP, Research Dept., MFA.

22.   William Menninger to Karl Menninger, Sep. 14, 1948; Karl Menninger to Milton Wexler, Nov. 22, 1950; Sibylle Escalona to William Menninger, [1949]; Sibylle Escalona to Milton Wexler, Oct. 8, 1948; all from SP, Research Dept., MFA.

23.   On Bleuler, see Ann Harrington, *Mind Fixers: Psychiatry's Troubled Search for the Biology of Mental Illness* (New York: Norton, 2019), 43–47. The term *schizophrenia* was also confusing, just as the term *dementia praecox* had been. Schizophrenia implied a "splitting" or "split personality," which no

experts thought was characteristic of the disease. Bleuler distinguished between "simple" symptoms such as disturbances in affect and in the association of ideas, "compound" symptoms such as disturbances in attention, and what he called "accessory" symptoms like hallucinations and delusions, which he thought were reactions to inner conflicts rather than defining symptoms of the illness. These were ideas that my father would later draw on in his own theory of schizophrenia.

24. Sigmund Freud, "The Defence Neuro-Psychoses," in *Freud: Early Psychoanalytic Writings* (1894; repr., New York: Collier, 1963), 67–82; Sigmund Freud, "Further Remarks on the Defence Neuro-Psychoses," in *Freud: Early Psychoanalytic Writings*, (1896; repr., New York: Collier, 1963), 151–74.

25. Sigmund Freud, "Psychoanalytic Notes Upon an Autobiographical Account of a Case of Paranoia (Dementia Paranoides)," in *Three Case Histories* (1911; repr., New York: Collier, 1963),161–66; Daniel Paul Schreber, *Memoirs of My Nervous Illness*, trans. and ed. Ida Macalpine and Richard A. Hunter (London: Dawson, 1955), 10–17.

26. Freud, "Psychoanalytic Notes," 173–76.

27. Freud used the terms *cathexis* or *libidinal cathexis* to indicate a charge of emotional energy, like a battery charge, positive or negative, directed at people or things in the real world or at mental representations and images; Sigmund Freud, "The Unconscious," in *General Psychological Theory: Papers on Metapsychology* (1915; repr., New York: Collier, 1963), 142–43; Sigmund Freud, "Reality in Neurosis and Psychosis," in *General Psychological Theory* (1924; repr., New York: Collier, 1963), 204; Sigmund Freud, "On Narcissism: An Introduction" in *General Psychological Theory* (1914; repr., New York: Collier, 1963), 57.

28. By interpreting the transference, the analyst theoretically reveals to their patient the often dysfunctional ways they relate to people, thereby enabling them to break free from the emotional baggage of the past and to respond more appropriately in the present. Freud argued that because they were no longer interested in the external world, some patients "are inaccessible to the influence of psychoanalysis and cannot be cured by our endeavours." Sigmund Freud, "A Note on the Unconscious in Psychoanalysis," in *General Psychological Theory*, 56–57. My father would later challenge aspects of the accepted understanding of transference, arguing that the analysand's responses were not always unrealistic or derived from the past.

29. Milton Wexler, "The Psychoanalytic Therapy of Schizophrenia: Research Project," May 25, 1950, SP, Research Dept., MFA.

30. Milton Wexler to Sibylle Escalona, Sep. 28, 1948; David Rapaport to Sibylle Escalona, Nov. 18, 1948; both from SP, Research Dept., MFA.

31. Milton Wexler, "Autobiographical Tape," Oct. 16, 1969. Fromm-Reichmann noted that "work with psychotics is obviously much more taxing for most psychiatrists than is therapy with neurotics." See Frieda Fromm-Reichmann, "Some Aspects of Psychoanalytic Therapy with Schizophrenics," in Brody and Redlich, *Psychotherapy with Schizophrenics*, 92.

2. OUT OF BROOKLYN

1. Alfred Kazin, *Walker in the City* (New York: Harcourt, Brace & Company, 1951), 169; Ann Fadiman, *The Wine-Lover's Daughter* (New York: Farrar, Strauss and Giroux, 2017), Apple Books.

2. Irving Lazar, *Swifty: My Life and Times* (New York: Simon & Schuster, 1995), 20.

3. Vivian Gornick, *The Men in My Life* (Cambridge, MA: MIT Press, 2008), 92.

4. Stephen Farber and Marc Green, *Hollywood on the Couch: A Candid Look at the Overheated Love Affair Between Psychiatrists and Moviemakers* (New York: William Morrow, 1993), 217.

5. Fellow psychologist, Brooklyn native, and later an acquaintance with whom he corresponded, Abraham Maslow, was born the same year as Dad and also recalled a childhood of frequent moves

but in a negative way, as contributing to his early shyness and difficulty making friends. See William L. Hoffman, *The Right to Be Human: A Biography of Abraham Maslow* (Los Angeles: Jeremy Tarcher, 1966), 5.

6. My father was never able to find his birth certificate. But while searching Ancestry.com after he died, I discovered a certificate for one "Samuel Wechsler" who was born August 24, 1908, to Nathan Wechsler and Mollie Skolnetsky, at 51 Portola Street, San Francisco, the exact place and date Dad remembered. Samuel Wechsler was Nathan's grandfather's name, so that may have been the first name given to Dad. Or possibly the clerk had made an error.

7. Henry Wexler, journal entry, May 22, 1969, Wexler family collection.

8. Henry Wexler, journal entry, Nov. 12, 1968, Wexler family collection.

9. Gerald Aronson, interview with the author, Sep. 2, 2014, Los Angeles, California.

10. Nathan Wexler's draft registration card of 1942 gives his birthplace as Odessa. However, family memory locates them in Kiev.

11. Balleisen was the partner more deeply in debt, with personal liabilities of about $15,000, while my great-grandfather owed about $5,000. More serious was their joint liability, for the firm claimed assets of $18,586 while their liabilities were nearly six times that amount, almost $120,000. I have no memory of hearing about Wolf Balleisen, possibly because he was an unsavory character, but it seems certain that this man was my great-grandfather's partner in various, possibly crooked, real estate deals in Brooklyn, as well as in a lawsuit in 1902, which the two men lost (involving a loan that apparently was never repaid) ("Goldstein Wins Suit," *Brooklyn Daily Eagle*, Mar. 8, 1902). In August 1907, the two men filed for bankruptcy, both individually and as partners, under the name of their company, Balleisen and Wexler. Events leading to the bankruptcy may have been another reason for my grandfather's move to San Francisco, for, according to the *Eagle*, "There are many creditors holding judgments and a number of unsecured creditors with claims for materials and services" ("Builders Bankrupt," *Brooklyn Daily Eagle*, Aug. 25, 1907). Morris must have applied for citizenship as soon as he was able, for he received his naturalization papers in 1892, thereby naturalizing the entire family.

12. In October 1908, Morris, along with his wife, Rose, and another couple, Robyn (or Reuben) and Rosa Grau, established another company, the Wexler-Grau Construction Company, incorporated in Brooklyn with a starting capital of $10,000 ("Brooklyn Corporations," *Brooklyn Daily Eagle*, Oct. 1, 1908). The following year, in 1909, Morris apparently decided to strike out on his own or else the company acquired a new name, the Wexler Realty and Construction Company of New York City, with two different partners this time, Max Hanbenstock and Jacob Tuck ("Brooklyn Investors," *Brooklyn Daily Eagle*, Nov. 1, 1909). This company evidently survived; see Milton Wexler, "Autobiography" (unpublished manuscript, 2005), Wexler family collection.

13. Henry Wexler to Milton Wexler, Nov. 18, 1962, Wexler family collection.

14. Henry Wexler, journal entry, Jul. 25, 1964, Wexler family collection. The year Dad (along with his siblings and Mollie) lived with their grandmother was a year in which Nathan and Mollie were separated; my father recalled that Mollie cried all the time, which may have been a response not only to her fear for the marriage but also a foreshadowing of her lifelong struggle with depression, another incentive for her sons' attraction to psychology. Far more than Dad, Henry mulled over the puzzle of their parents' lives. "What is the truth about Mother?" Henry wrote in his journal soon after Nathan died. "And how can one convey it?" (Nov. 21, 1962). Dad had fewer contacts with his East Coast relatives after 1951 and remembered the Skolnicks more benignly than Henry did. He was proud of the family organization, the Resnikoff Foundation, a mutual aid society that he said he had helped organize in the 1930s to assist the poorer relatives. A photograph from 1935 shows a gathering of foundation members (with at least one hundred people present) in a large restaurant. Dad, who was then in his twenties and already an attorney, sits in the back sporting a mustache and a quizzical expression, a few feet behind Boris Thomashevsky, a famous actor in the Yiddish theater who was distantly related to Mollie by marriage.

15. *RVM*, 12.

16. "With Brooklyn and Long Island Students in School and College," *Brooklyn Daily Eagle*, Apr. 4, 1925; "Commercial High Debating Team Champions," *Brooklyn Daily Eagle*, Apr. 15, 1924.

17. "Boys High Defeats Erasmus in Debates; Eastern District H.S. Wins Over Alexander Hamilton," *Brooklyn Daily Eagle*, Mar. 28, 1925; Milton Wexler, auto. tape, 1969. He would later title a chapter of "Winning Ways," a self-help manuscript, "I Could Have if I Would Have" to emphasize the point.

18. "Forensic Scholarship Won by Brooklyn Boy," *Brooklyn Daily Eagle*, Apr. 4, 1925; "Many Prizes Given at Hamilton High Class Exercises," *Brooklyn Daily Eagle*, Jun. 10, 1925; "Class President Will Enter Harvard in Fall," *Brooklyn Daily Eagle*, Jun. 26, 1925. The statistically minded school principal noted in *The Ledger* that 172 members of the graduating class, about three-quarters of the pupils, "had some business experience during summer vacation or after school" and that about half, 124 students, partly supported themselves during the school year. Nine supported themselves "entirely" (*The Ledger*, Senior Number, vol. XXI, no. 5). The Commercial High School (later Alexander Hamilton High) student body consisted of the sons of small business owners and working-class parents who expected their teenaged children to help support themselves and whose professional ambitions were limited. It is not surprising, then, that while students voted English the most valuable subject, bookkeeping and accounting came in second. In fact, sixty-five graduating seniors, about 30 percent, expected to take up accounting as a profession, while much smaller numbers headed for law (thirty-three, about 15 percent), teaching (sixteen, or 7 percent), business (fifteen, about 7 percent), medicine (only six, about 3 percent), and dentistry (five, or 2 percent). Perhaps the bar was relatively low—of the 1,065 boys who had entered Commercial High, only 222 graduated four years later. This Brooklyn high school with a predominantly Jewish student body had a surprisingly low graduation rate: only 21 percent of enrolled pupils graduated four years later, suggesting that the majority were sons of working-class parents who needed them to get jobs to bring in an income. But of the 222 who did graduate, more than 90 percent expected to attend college, most planning on evening classes because they needed to work during the day. Most also had summer jobs.

19. Milton Wexler, "Valedictory Speech, June 1925," author's collection.

20. Milton Wexler, transcript, Syracuse University, Syracuse, New York; Milton Wexler, transcript, Teachers College, Columbia University, New York, New York. Dad received three degrees from Teachers College: a BS in June 1939, an MA in June 1940, and a PhD in January 1949.

21. Gornick, *The Men in My Life*, 93–99.

22. He apparently earned a little money selling insurance, for after his death I discovered among his papers a small business card with "Wexler and Starobin, Insurance Brokers," printed on it, though he never mentioned this enterprise to Nancy or me.

23. Milton Wexler, auto. tape, 1969.

24. These friends included Saul Machover and Bert Wiggers, an analyst who later came to Menninger.

25. As Eli Zaretsky has described, early twentieth-century psychoanalysis had an antiestablishment countercultural cast. A number of eminent analysts, including some close to Freud, had backgrounds in law, journalism, politics, art, and even music, rather than medicine. See Eli Zaretsky, *Secrets of the Soul: A Social and Cultural History of Psychoanalysis* (New York: Vintage, 2005), 7.

26. Between 1927 and 1930, perhaps forty American psychiatrists went abroad for psychoanalytic training. See Nathan Hale, *The Rise and Crisis of Psychoanalysis in the United States: Freud and the Americans, 1917–1985* (New York: Oxford University Press, 1995), 114.

27. That professor was the New York State Supreme Court judge Joseph Force Crater, who vanished one day in 1930 and was never heard from again.

28. "Three Brooklyn Students Win Law Scholarships," *Brooklyn Daily Eagle*, Aug. 30, 1928; "851 Students Pass Examinations; Certificates Due," *Brooklyn Daily Eagle*, Aug. 23, 1930.

29. *RVM*, 104.

30. *RVM*, 26.

3. BECOMING FREUDIAN

1. I am grateful to Tulle Hazelrig for the gift of a rare offprint by Henry E. Crampton, "A History of the Department of Zoology of Columbia University," reprinted from *Bios* 21, no. 4 (December 1960); Leonore Sabin to Mrs. L., n.d., author's collection.

2. On Leonore Sabin's honors, see Hunter College yearbook, *Wisterion, 1934* (New York: Schilling, 1934), 86, 162. See also Leonore Sabin to Mrs. L., n.d., author's collection.

3. On the early history of Huntington's see Alice Wexler, *The Woman who Walked into the Sea: Huntington's and the Making of a Genetic Disease* (New Haven, CT: Yale University Press, 2008); on Nazi policies see Benno Muller-Hill, *Murderous Science: Elimination by Scientific Selection of Jews, Gypsies, and Others in Germany, 1933–1945* (Cold Spring Harbor, NY: Cold Spring Harbor Laboratory, 1998).

4. "400,000 Germans to Be Sterilized," *New York Times*, Dec. 21, 1933; "Sterilization Law Is Termed Humane," *New York Times*, Jan. 22, 1934; C. C. Hurst, "Germany's Sterilization Law: What It Might Accomplish," *New York Times*, Aug. 5, 1934; "Eugenic Exposition Opened in Germany," *New York Times*, Mar. 23, 1935; "Nuptial Health Decreed. German Law Establishes Disease as Bar to Marriage," *New York Times*, Oct. 19, 1935; "Curbing Disease: Voluntary Sterilization Found Beneficial," *New York Times*, Mar. 8, 1936.

5. The literature on eugenics is vast. On eugenics in the United States, see for example Diane Paul, *Controlling Human Heredity: 1865 to the Present* (Amherst, NY: Humanity, 1998); Alexandra Minna Stern, *Eugenic Nation: Faults and Frontiers of Better Breeding in Modern America* (Berkeley: University of California Press, 2005); Daniel J. Kevles, *In the Name of Eugenics: Genetics and the Uses of Human Heredity* (Berkeley: University of California Press, 1985). See also the Image Archive of the American Eugenics Movement (Dolan DNA Learning Center, Cold Spring Harbor Laboratory), available at www.eugenicsarchive.org, for an online collection of images and essays.

6. See Paul, *Controlling Human Heredity*, 117–21.

7. Paul Popenoe, "The German Sterilization Law," *Journal of Heredity* 25, no. 7 (Jul. 1934): 257–60.

8. Leonore Wexler, notes on scholastic and teaching attainments, to Mrs. L., n.d., author's collection.

9. I am grateful to Jackie Wehmueller for this insight.

10. *Civilization and Its Discontents* was published (in German) in 1930, *New Introductory Lectures* in 1933, and *The Problem of Anxiety* in 1936. *Basic Writings of Sigmund Freud* was published in this country as a Modern Library Giant in 1938. The *Times* regularly noted Freud's honors, such as the Goethe Prize, his election to the Royal Society in London, and his inclusion on lists of "The Ten Greatest Living Jews." "Picks Ten Greatest Jews," *New York Times*, Apr. 6, 1936.

11. The New York Psychoanalytic Society and Institute, formed in 1931, was soon to become the most influential center for psychoanalytic training in the country. Nathan Hale, *The Rise and Crisis of Psychoanalysis in the United States: Freud and the Americans, 1917–1985* (New York: Oxford University Press, 1995), 189.

12. In 1926, the New York state legislature had outlawed the practice of medicine by anyone who was not a registered medical doctor and forbade anyone without a medical degree to call him- or herself "doctor" if they were involved in "public health" or the diagnosis and treatment of physical illness. Hale, *The Rise and Crisis of Psychoanalysis*, 33.

13. Reik's doctoral thesis at the University of Vienna, one of the earliest psychoanalytic studies of literature after Freud's, was an interpretation of Flaubert's novella *The Temptation of Saint Anthony*.

14. "American Loses Suit Against Freud," *New York Times*, May 25, 1927.

15. Freud believed that medical training could be a handicap for analysts and persuaded Reik to pursue an advanced degree in the humanities and not medical school, as he had intended. Peter Gay writes, "All his life Freud was intent on preserving the independence of psychoanalysis from the doctors no less than from the analysts." Peter Gay, *Freud: A Life for Our Time* (New York: Norton, 1988), 492.

16. "Average U.S. Income Told: Earnings of Typical American Family in 1938 Put at $2,116," *California and Western Medicine* 50, no. 5 (May 1939): 389. In 1938, $2,000 was equivalent to about $36,000 in 2020. The average annual income of New Yorkers in 1938 was estimated to be $3,069.

17. Lawrence Cremin, *A History of Teachers College, Columbia University* (New York: Columbia University Press, 1954), 164; "Topics of the Times," *New York Times*, Feb. 25, 1937.

18. Milton Wexler to Alice Wexler and Nancy Wexler, Feb. or Mar. 1965, author's collection.

19. Milton Wexler, auto. tape, 1969; Milton Wexler, "Autobiography" (unpublished manuscript, 2005), Wexler family collection. My father's experience at Teachers College contrasted dramatically with that of Maslow at the New School, who encountered many distinguished European intellectuals in flight from the Nazis, most of whom were accessible and welcoming to students like himself. Unlike my father, Maslow felt he had had "the best teachers, both formal and informal, of any person who ever lived, just because of the historical accident of being in New York City when the very cream of European intellect was migrating away from Hitler. New York City in those days was simply fantastic." One reason for my father's lack of interest in the New School may have been the presence there of dissident psychoanalysts, such as Alfred Adler, Karen Horney, and Erich Fromm, while he was focused more narrowly on Freud and on getting the professional credential of a PhD. See Edward Hoffman, *The Right to Be Human: A Biography of Abraham Maslow* (New York: J. P. Tarcher, 1988), 86-87.

20. Nonetheless, in his published PhD dissertation on an entirely different topic, Dad thanked Symonds effusively, noting that his ability to complete the dissertation was "entirely due to the quiet confidence and helpful advice he offered in liberal quantities at every stage of its progress. As teacher, critic, and therapist for the anxieties of authorship he has shown incomparable skill." Milton Wexler, reference material, "An Investigation of Psychiatric Screening Tests in the United States Navy" (PhD diss., Teachers College, Columbia University, 1948).

21. Milton Wexler, "Case History of a 'Bad' Boy" (unpublished manuscript, ca. 1942), author's collection. Dad later listed "The Story of Frank" as a publication on one of his CVs, stating that it was published in the *Journal of Exceptional Children* in "approximately 1942," though I have been unable to locate it.

22. Percival M. Symonds, *Adolescent Fantasy: An Investigation of the Picture-Study Method of Personality Study* (New York: Columbia University Press, 1949), 271.

23. Percival M. Symonds, *Dynamics of Psychotherapy: The Psychology of Personality Change*, vol. 1 (New York: Grune & Stratton, 1956), x–xi.

24. Milton Wexler to Alice Wexler and Nancy Wexler, Oct. 9, 1965, author's collection.

25. Milton Wexler, auto. tape, 1969.

26. Milton Wexler, "Tribute and Contribution," in *Explorations in Psychoanalysis: Essays in Honor of Theodor Reik, on the Occasion of His Sixty-Fifth Birthday, May 12, 1953*, ed. Robert Lindner (New York: Julian, 1953), 157; Milton Wexler, "Tribute and Contribution" (unpublished manuscript, 1953), author's collection.

27. Milton Wexler, "Tribute and Contribution," in Lindner, *Explorations in Psychoanalysis*; Theodor Reik, *Listening with the Third Ear* (New York: Grove, 1948), 312, 438.

28. Milton Wexler, "Tribute and Contribution," in Lindner, *Explorations in Psychoanalysis*.

29. Dany Nobus, "*Fluctuat Nec Mergitur* or What Happened to Reikian Psychoanalysis?" *Psychoanalytic Psychology* 23, no. 4 (2006): 693. According to Nobus, Lacan "borrowed heavily from Reik's critique in *Listening with the Third Ear*."

30. Milton Wexler, auto. tape, 1969.

31. Milton Wexler, "Psychoanalysis as a Basis for Psychotherapeutic Training," in *New Horizon for Psychotherapy: Autonomy as a Profession*, ed. Robert R. Holt (New York: International Universities Press, 1971), 189–90.

32. Milton Wexler to attorney general, February 20, 1940; Robert R. Hannon to Milton Wexler, Mar. 5, 1940; John H. Bennett to Milton Wexler, Mar. 11, 1940; Robert R. Hannon to Milton Wexler, Mar. 13, 1940; all from author's collection.

33. With many publications, years of practice, and his reputation as a respected colleague of Freud, a member of his inner circle even, Reik expected to be welcomed warmly by the American analysts in New York. Instead, arriving around the time the American Psychoanalytic Association passed the 1938 rule, he found himself treated dismissively. The New York analysts offered to pay him a stipend if he would limit himself to psychoanalytic teaching, research, and writing, a demeaning offer that would hardly have enabled him to support his family. Although Reik did eventually give seminars at the New York Institute and practiced analysis without interference, he was never accepted as a regular member, which may have been one of the reasons he was so welcoming to people like my father. Reik remained bitter about his exclusion from the mainstream of American analysis, eventually forming his own institute, the National Psychological Association for Psychoanalysis, which offered training to people without medical degrees. It continues to this day. See Murray H. Sherman, "Theodor Reik and Lay Analysis," *Psychoanalytic Review* 75, no. 3 (Fall 1988): 380–91; Martin A. Schulman, "The Question of Lay Analysis Reconsidered," *Psychoanalytic Psychology* 23, no. 4 (2006): 701–10.

34. Dad expressed little interest in Reik's National Psychological Association for Psychoanalysis, which Reik established in 1948. But as Dany Nobus has written, "One can only be a Reikian by not being a follower, which also entails not being a follower of Reik." Nobus, "*Fluctuat*," 696.

35. Milton Wexler, "A Look Through the Rearview Mirror" (unpublished manuscript, 2002), author's collection. Dad also thought Reik became too much of a popularizer, writing too much and too superficially in an effort to become better known than his critics.

36. Milton Wexler, "Beyond Conflict" (lecture, "Psychoanalysis and the Image of Man" conference in honor of the seventieth birthday of Theodor Reik, New York, May 18, 1953): "Is it any wonder then that the popularity of most neo-psychoanalytic movements stems from their efforts to attenuate the significance accorded to instincts and to man's biological nature?" But see also Dagmar Herzog, *Cold War Freud: Psychoanalysis in an Age of Catastrophe* (Cambridge: Cambridge University Press, 2017), Apple Books.

37. Milton Wexler, "Beyond Conflict"; "Some Virtue Seen for Delinquency," *New York Times*, May 19, 1958.

38. Herzog, *Cold War Freud*, part 1, "Leaving the World Outside." See also Eli Zaretsky, *Secrets of the Soul: A Social and Cultural History of Psychoanalysis* (New York: Vintage, 2004), 176–77. Yet, Zaretsky writes, when many of the most profound thinkers of the fifties, such as Norman O. Brown and Herbert Marcuse, sought to criticize that Cold War ethos, they turned to psychoanalytic theory for help. See Zaretsky, *Secrets of the Soul*, 316–320; also *Political Freud: A History* (New York: Columbia University Press, 2015).

39. Milton Wexler, autobiog., (unpublished manuscript, 2005), author's collection.

40. Milton Wexler, auto. tape, 1969.

41. When the father of a former patient whom he had "cured" of homosexuality called on him in Washington to express his gratitude, Dad decided to ask this man, an admiral in the Navy, to help him get into the Navy. Milton Wexler, auto. tape, 1969.

42. Milton Wexler, auto. tape, 2005; Milton Wexler, "Constructing and Using Achievement Tests," Bureau of Naval Personnel, Test and Research Unit, [1943], Washington DC. This paper was eventually published as a chapter, "Measures of Personal Adjustment," in *Personnel Research and Test Development in the Bureau of Naval Personnel*, ed. Dewey B. Stuit (Princeton, NJ: Princeton University Press, 1947), 174. In November 1945, Dad was on a panel presenting his wartime research at a conference of the Military Division of the American Psychological Association, suggesting his early participation in professional activities of psychologists. See "Psychiatric Screening of Naval Personnel," in *New Methods in Applied Psychology: Proceedings, Maryland Conference on Military Psychology*, ed. G.A. Kelly (College Park, MD: University of Maryland Press, 1947), 60–66.

43. *RVM*, 136–37; see also Irwin Rosen, "Psychoanalysis at the Menninger Foundation: Beginnings, Growth, and National Influence," Oral History Workshop #4, San Juan, Puerto Rico, May 7, 1981.

4. THE SLAP, EXPLAINED

1. Irwin Rosen, "Psychoanalysis at the Menninger Foundation: Beginnings, Growth, and National Influence," Oral History Workshop #14, San Juan, Puerto Rico, May 7, 1981, 26; on the personalities at Menninger, see Lawrence J. Friedman, *Menninger: The Family and the Clinic* (New York: Knopf, 1990).

2. Friedman, *Menninger*, 180–82.

3. Maimon (Mike) Leavitt, interview with the author, Sep. 14, 2000, Los Angeles; the term *clinical conquistador* can be found in Gerald Aronson, "What Was It Like in Topeka?" *Psychohistory Review* 19, no. 1 (1990): 145.

4. Herbert Schlesinger, interview with the author, Jul. 23, 2014, New York.

5. Gerald Aronson, interview with the author, Sep. 2, 2014, New York.

6. Milton Wexler, auto. tape, 1969.

7. Frieda Fromm-Reichmann, Gertrud Schwing, and Marguerite Sechehaye for instance.

8. Milton Wexler, "The Structural Problem in Schizophrenia: Therapeutic Implications" (unpublished manuscript presented at the June 24, 1950, meeting of the Topeka Psychoanalytic Society), author's collection.

9. Karl Menninger to Milton Wexler, Jun. 26, 1950, SP, Research Dept., MFA.

10. Trygve Braatøy to Milton Wexler, [1950]; Trygve Braatøy to Karl Menninger, Jun. 30, 1950; both from SP, Research Dept., MFA.

11. Milton Wexler, "The Structural Problem in Schizophrenia: Therapeutic Implications," *International Journal of Psychoanalysis* 32, no. 3 (1951): 162.

12. Freud proposed that the superego is formed from the child's internalizing or identifying with a parent and his or her moral strictures, manifested through patterns of approval and disapproval, reward and punishment. In Nedda's case, "Between us and the ego which has abdicated its legitimate function of reality testing lies a turbulent and chaotic battleground where instinctual impulses of many varieties threaten constant eruption even in the face of unremitting superego pressures which threaten annihilation as the measure to be paid for the least gratification of such impulses." Milton Wexler, "The Structural Problem," 160.

13. *RVM*, 204–5. Recordings and transcripts of my father's sessions with Nedda may be stored in the patient records in the Menninger Clinic Medical Records Department at the Baylor College of Medicine in Houston. However, I was unable to gain access to them.

14. Robert Wallerstein, interview with the author, Sep. 8, 2014, Belvedere, California.

15. "There would be no pressure on the ego-organization to undertake aggressive or adaptive behavior. Instead, just as in an organic illness, the libido would be drawn away from its attachments in the ego-organization in order to neutralize the mortido. These phenomena constitute what I conceive to be the pathogenic process in schizophrenia." William L. Pious, "The Pathogenic Process in Schizophrenia," *Bulletin of the Menninger Clinic* 13, no. 5 (Sep. 1949), 154–57.

16. Pious, "The Pathogenic Process," 156.

17. Gail A. Hornstein, *To Redeem One Person Is to Redeem the World: The Life of Frieda Fromm-Reichmann* (New York: Other, 2000), 133–35.

18. See, for instance, Mari Jo Buhle, *Feminism and Its Discontents* (Cambridge, MA: Harvard University Press, 1998); Ferdinand Lundberg and Marynia F. Farnham, *Modern Woman: The Lost Sex* (New York: Harper & Brothers, 1947); Silvano Arieti, *Interpretation of Schizophrenia* (New York: International Universities Press, 1974). Dad also cited John Rosen's claim that "with chronic schizophrenics . . . the most important etiological agent is an unconsciously hostile mother" (Pious, "The Pathogenic Process," 157). Neither my father nor Pious presented any evidence that their patients' mothers had been hostile or destructive, nor did they seem interested in these mothers at all.

19. Milton Wexler, "Principles of Intensive Psychotherapy by Dr. Frieda Fromm-Reichmann" (unpublished manuscript, [1950]), author's collection.

20. Frieda Fromm-Reichmann to Milton Wexler, Nov. 10, 1953, author's collection. Hornstein notes that in her Yale paper, Fromm-Reichmann revised her earlier views about treating schizophrenia, emphasizing the need for boundaries and for something other than "indirectness and sweetness," views that accorded closely with those of my father. See Hornstein, *To Redeem One Person*, 302–3. My father's harsh critique was not limited to Fromm-Reichmann. His criticisms as a discussant were sometimes so severe that the recipient occasionally responded in kind. "Dr. Wexler's comments are of a different order than those of the other discussants," wrote L. Bryce Boyer in response to one of Dad's assaults. "He seems to have attempted to discredit my work through the use of distortion and ridicule and by inferring that I am either naive or dishonest." Indeed, Dad sometimes noted that his colleagues expressed concern regarding his harsh tone, to which he replied that they seemed to care more about political expediency than about "any vehement attempt to arrive at scientific truth." L. Bryce Boyer, "Office Treatment of Schizophrenic Patients by Psychoanalysis," *Psychoanalytic Forum* 1, no. 4 (Winter 1966), 337–66.

21. Milton Wexler, "The Structural Problem in Schizophrenia: The Role of the Internal Object," *Bulletin of the Menninger Clinic* 15, no. 6 (Nov. 1951), 221–34; also in Eugene B. Brody and Frederick C. Redlich, eds., *Psychotherapy with Schizophrenics: A Symposium* (New York: International Universities Press, 1952), 179–201.

22. Milton Wexler, "The Structural Problem in Schizophrenia: The Role of the Internal Object," in *Bulletin of the Menninger Clinic*, 221–34.

23. Karl Menninger, "Memo from Dr. Karl Menninger," n.d., SP, Research Dept., MFA; Wexler, "The Structural Problem in Schizophrenia: The Role of the Internal Object," in *Bulletin of the Menninger Clinic*; Brody and Redlich, *Psychotherapy with Schizophrenics*, 199–200.

24. Robert Bak, "Discussion of Dr. Wexler's Paper," in Brody and Redlich, *Psychotherapy with Schizophrenics*, 202–7; Ludwig Eidelberg, "Discussion of Dr. Wexler's Paper," in Brody and Redlich, *Psychotherapy with Schizophrenics*, 208–15. Silvano Arieti claimed that "Wexler feels that the therapist should be harsh and strict, and should assume a generally repressive role." Arieti, *Interpretation of Schizophrenia*, (1974; Chevy Chase, MD: International Psychotherapy Institute, e-book 2015), 1037–38.

25. Monthly report, Apr. 1951, SP, Research Dept., MFA.

26. Milton Wexler, "Discussion of 'Remarks to an Experiment in the Treatment of Schizophrenic Patients: The Use of Psychoanalysis with Minimal Parameters' by L. Bryce Boyer" (paper presented, San Francisco, Jan. 16, 1964), author's collection. See also Milton Wexler, "Discussion of 'Office Treatment of Schizophrenic Patients by Psychoanalysis,' by L. Bryce Boyer, M.D.," *Psychoanalytic Forum* 1, no. 4 (Winter 1966), 347–49.

27. Karl Menninger to Milton Wexler, Aug. 24, 1989, author's collection.

5. EX-TOPEKAN

1. Milton Wexler to David Rapaport, Jan. 30, 1951, DRP.

2. Morris Bender to David Rubenfine, Sep. 30, 1950, author's collection.

3. At some point Karl Menninger and probably others besides Bernstein must have learned about the presence of Huntington's chorea in Mom's family. See Karl Menninger to Milton Wexler, Aug. 24, 1989: "Yes, I remember the sad anticipations we had about your wife's family affliction before you left here. She and her brothers are all gone now, you say. Sad."

4. Peggy Papp, "The Worm in the Bud: Secrets Between Parents and Children." In *Secrets in Families and Family Therapy*, ed. Evan Imber-Black (New York: Norton, 1993), 67–85. Years later, in his memoir, Dad claimed that his brother Henry, who had a medical degree, did research on Huntington's

chorea in 1950, at the time of Mom's brothers' diagnoses, and informed him and Mom that the disease affected only men. Dad said that he subsequently felt relieved and decided to focus solely on the financial support of Mom's affected brothers (*RVM*, 212). A few years later, when Dad, Nancy, and I discussed this issue—and the facts that even George Huntington had described an affected woman and that no reputable medical literature claimed that Huntington's was sex linked—Dad allowed that if he had been told the truth, he did not absorb it. "From that moment on, it felt very comfortable in my own family. I cannot recall any sense of threat, any notion that I must observe my wife, question her history, worry about my two beloved children." Milton Wexler, "Autobiography" (unpublished manuscript, 2005), 44, author's collection.

5.   Milton Wexler, auto. tape, 1969.

6.   *RVM*, 216.

7.   Irwin Rosen, "Analysis at the Menninger Foundation: Beginnings, Growth, and National Influence," from Oral History Workshop #14, San Juan, Puerto Rico, May 7, 1981. Boston Psychoanalytic Society and Institute, Oral History Transcripts, 1961–2018, Box 6.

8.   David Rapaport to Milton Wexler, Jun. 12, 1951, DRP.

9.   Milton Wexler to David Rapaport, Jun. 19, 1951, DRP.

10. Milton Wexler to David Rapaport, Jun. 5, 1951, DRP.

11. Larry Ceplair and Steven Englund, *The Inquisition in Hollywood: Politics in the Film Community, 1930–60* (Champaign: University of Illinois Press, 2003), 362–63.

12. Some of the most eminent and independently minded Los Angeles analysts joined the Institute for Psychoanalytic Medicine of Southern California; for example, Judd Marmor, who helped remove homosexuality from the American Psychiatric Association's *Diagnostic and Statistical Manual of Mental Disorders*, and Robert Stoller, a pioneer in rethinking sexual fantasies and gender identity. See also Douglas Kirsner, *Unfree Associations: Inside Psychoanalytic Institutes* (London: Process, 2000), 140–42.

13. Leonore Wexler to Aline Mintz, 1952, author's collection.

14. "Film Actor Praises Menninger Clinic," *Los Angeles Times*, [Jun. 18, 1949].

15. Jean Stein, *West of Eden: An American Place* (New York: Random House, 2016), 181.

16. Hedda Hopper, "Film Work Resumed by Walker," *Los Angeles Times*, Aug. 28, 1949.

17. Hedda Hopper, "New Robert Walker Returns to Studio," *Los Angeles Times*, July 11, 1949.

18. "Death Takes Actor Robert Walker," *Los Angeles Times*, Aug. 29, 1951; see also Stephen Farber and Marc Green, *Hollywood on the Couch: A Candid Look at the Overheated Love Affair Between Psychiatrists and Moviemakers* (New York: William Morrow, 1993), 115–16.

19. Milton Wexler to David Rapaport, Oct. 5, 1951, DRP.

20. While he considered himself a classic Freudian, the titles of some of his papers show a certain poetic flair; for example, "The Mother Tongue and the Mother" (1950), "On Boredom" (1953), "Empathy and Its Vicissitudes" (1960), and "On the Silence and Sounds of the Analytic Hour" (1961).

21. Milton Wexler to David Rapaport, Apr. 22, 1953, DRP.

22. Milton Wexler to David Rapaport, Mar. 20, 1954, DRP.

23. David Rapaport to Milton Wexler, Mar. 25, 1954, DRP.

24. Milton Wexler to David Rapaport, 1954, DRP.

25. David Rapaport to Milton Wexler, Apr. 13, 1954, DRP.

26. Milton Wexler to David Rapaport, Apr. 1956; David Rapaport to Milton Wexler, Apr. 23, 1956; both from DRP.

27. David Rapaport to Milton Wexler, Sep. 29, 1955, DRP.

28. Milton Wexler, "Notes on Psychotherapy in Schizophrenia" (lecture presented at a meeting of the Southern California branch of the American Psychiatric Association, Los Angeles, Feb. 6, 1953).

29. Since some time had elapsed since their last discussion of these issues, Dad reminded Rapaport that "my clinical experiences have made me very interested in the problems connected with the

internalization of objects and in the complex internal relationships with these objects. (Not yet in the Fairbairn manner.)" Milton Wexler to David Rapaport, Apr. 1955, DRP.

30.   Milton Wexler to David Rapaport, Mar. or Apr. 1955, DRP.

31.   William L. Pious, "Obsessive-Compulsive Symptoms in an Incipient Schizophrenic," *International Journal of Psychoanalysis* 32 (1951): 330; William L. Pious, "The Pathogenic Process in Schizophrenia," *Bulletin of the Menninger Clinic* 13, no. 5 (Sep. 1949), 152–59. See also Thomas H. McGlashan, "Psychosis as a Disorder of Reduced Cathectic Capacity: Freud's Analysis of the Schreber Case Revisited," *Schizophrenia Bulletin* 35, no. 3 (2009): 479. According to McGlashan, "Deficit has been prominent in most labels and descriptions of psychosis through the twentieth century."

32.   Milton Wexler to David Rapaport, Mar. or Apr. 1955, DRP.

33.   David Rapaport to Milton Wexler, Apr. 12, 1955; Milton Wexler to David Rapaport, Apr. 18, 1955; both from DRP.

34.   Milton Wexler to David Rapaport, Aug. 14, 1959, DRP.

35.   David Rapaport to Milton Wexler, Sep. 21, 1959, DRP.

36.   Milton Wexler to Henry Wexler, May 1, 1957, Wexler family collection.

6. LOSING THE ROAD MAP

1.   Milton Wexler to David Rapaport, Apr. 18, 1955, DRP.

2.   Milton Wexler to David Rapaport, May 21, 1958, DRP.

3.   Milton Wexler to David Rapaport, Feb. 5, 1957, DRP.

4.   Robert Wallerstein, interview with the author, Belvedere, California, Jun. 28, 2014; Robert Holt to Alice Wexler, email communication, Jun. 25, 2014.

5.   Karl Menninger to Milton Wexler, Aug. 23, 1978, author's collection; Milton Wexler to "Dearest Pooch" (Nancy Wexler), Oct. 18, 1971, Nancy Wexler collection; Alfred Goldberg, interview with the author, Los Angeles, Oct. 2, 2014.

6.   Philip R. A. May, *Treatment of Schizophrenia: A Comparative Study of Five Treatment Methods* (New York: Science House, 1968), 51.

7.   May, *Treatment of Schizophrenia*, 39–41. There were also four computer analysts, eleven social scientists, a biologist, a biostatistician, and four volunteers along with an office staff of thirty-three, and May, the principal investigator.

8.   May, *Treatment of Schizophrenia*, 232.

9.   May argued that my father's approach could be considered object relations therapy as opposed to interpretation-insight therapy, the more classically psychoanalytic approach. Dad did not refer specifically to object relations theorists in his writing and did not credit Klein or Fairbairn in his papers. He criticized ego psychology for downplaying the power of the id (drives) despite his ardent admiration for Rapaport.

10. Milton Wexler, "One Path to Recovery Lies in the Development of Stable Object Relationships with Someone in the Outside World," in May, *Treatment of Schizophrenia*, 282. My father's concerns were justified. The Camarillo study was later lauded as "revolutionary," the first "to prove the superiority of the drug phenothiazine" over psychotherapy in the treatment of schizophrenia. "Dr. Philip May, Noted for Studies of Schizophrenia, Dies," *LA Times*, Dec. 14, 1986; Fred Alvarez, "Renowned Research Unit a Study in Closure," *LA Times*, Feb. 16, 1997. The apparent success of drugs in treating the agitation, aggression, and anxiety of those with a diagnosis of schizophrenia along with inadequate funding for community mental health centers and new discoveries in neuroscience and genetics all contributed to increased psychiatric focus on medication and downplaying of psychotherapy. For a more positive assessment of psychotherapy

combined with medication and social support see Benedict Cary, "New Approach Advised to Treat Schizophrenia," *New York Times*, Oct. 20, 2015.

11. Joel T. Braslow, "Psychosis Without Meaning: Creating Modern Clinical Psychiatry, 1950 to 1980," *Culture, Medicine, and Psychiatry* 45 (2021): 429–55; Milton Wexler, "Hypotheses Concerning Ego Deficiency in Schizophrenia," in *The Out-patient Treatment of Schizophrenia: A Symposium*, ed. Sam C. Scher and Howard R. Davis (New York: Grune & Stratton, 1960), 33–43. This paper was presented at a symposium at the University of Minnesota. See also Milton Wexler, "Schizophrenia: Conflict and Deficiency," *Psychoanalytic Quarterly* 11, no. 1 (1971): 83–99. This paper was presented at a NIMH working group meeting on schizophrenia.

12. Wexler, "Hypotheses Concerning Ego Deficiency," 40.

13. Wexler, "Schizophrenia: Conflict and Deficiency," 83–84.

14. Wexler, "Schizophrenia: Conflict and Deficiency," 94.

15. Milton Wexler, "Discussion of 'The Non-transference Relationship in the Psychoanalytic Situation,'" *International Journal of Psychoanalysis* 51, no. 2 (1970): 148.

16. Ralph R. Greenson and Milton Wexler, "The Non-transference Relationship in the Psychoanalytic Situation," *International Journal of Psychoanalysis* 50, no. 1 (1969): 34.

17. Milton Wexler to "Dearest Ali" (Alice Wexler), 1969, author's collection.

18. Mari Jo Buhle, *Feminism and Its Discontents: A Century of Struggle with Psychoanalysis* (Cambridge, MA: Harvard University Press, 1998). Buhle writes that "most psychoanalysts endorsed the patriarchal family as the American ideal," blaming feminism for disrupting the "normal" division of labor between the sexes (194–95). But she also points out that feminist psychoanalysts and psychoanalytic feminists used psychoanalytic concepts to critique patriarchal ideology. As Juliet Mitchell famously wrote, "However it may have been used, psychoanalysis is not a recommendation *for* a patriarchal society, but an analysis *of* one." Juliet Mitchell, *Psychoanalysis and Feminism: Freud, Reich, Laing, and Women* (New York: Vintage, 1975), xiii.

19. See Dagmar Herzog, *Cold War Freud: Psychoanalysis in an Age of Catastrophe* (Cambridge: Cambridge University Press, 2017), Apple Books.

20. Ann Harrington, *Mind Fixers: Psychiatry's Troubled Search for the Biology of Mental Illness* (New York: Norton, 2019); Joel T. Braslow, "Psychosis Without Meaning: Creating Modern Clinical Psychiatry, 1950 to 1980," *Culture, Medicine, and Psychiatry* 45 (2021): 429–55.

21. Frederick C. Appel, "Experts Disagree on a Worm's I.Q.," *New York Times*, Feb. 14, 1965; Marcia Meldrum, Joel Braslow, and Rena Selya, *A History of the Society for Neuroscience* (Washington, DC: Society for Neuroscience, 2021), http://www.sfn.org/about/history-of-sfn/1969-2019; Eric R. Kandel, *In Search of Memory: The Emergence of a New Science of Mind* (New York: Norton, 2006), esp. 363–75.

22. Kandel, *In Search of Memory*, 366–67. Dad would have agreed with the Nobel prize–winning neuroscientist Eric Kandel that what he calls the new biology of mind will be richer and more meaningful if reached through a synthesis with psychoanalysis. See also Eric Kandel, *The Disordered Mind: What Unusual Minds Tell Us About Ourselves* (New York: Farrar Straus, 2018), 248. On "many workers" who looked to physiology for answers, see Robert P. Knight, introduction to *Psychotherapy with Schizophrenics*, ed. Eugene B. Brody and Fredrick C. Redlich (New York: International Universities Press, 1952), 11; Meldrum et al., *History of the Society*, 47–49; Harrington, *Mind Fixers*. My father also recalled discussions at Menninger in which "the European analysts insisted that there were reactive and process schizophrenics, some with organic features and some with severe reactions to environmental pressures." Milton Wexler, "Autobiography" (unpublished manuscript, 2005), 518–19, author's collection. A few years after Rome, already immersed in Huntington's disease research, my father became intrigued by the speculation of a prominent neuroscientist, Eugene Roberts, that a disturbance in the neurotransmitter GABA might be involved in schizophrenia. He wanted to know if Roberts's speculations concerning GABA and the inhibitory neurons had any relation to issues of memory, reminding Roberts of Freud's claim that "the process of reality testing is not one

of discovery but of rediscovery," an idea to which Dad often referred. (Milton Wexler to Eugene Roberts, Sep. 27, 1971, author's collection). But by 1971 he was not in a position to pursue this theory although he would develop a relationship with Roberts and with a younger neuroscientist, Steven Matthysse, who also saw correspondences between the two diseases.

23.   Jonathan M. Metzl, *The Protest Psychosis: How Schizophrenia Became a Black Disease* (Boston: Beacon, 2011), xiii–xxi.

24.   Eric Kandel reported that drugs were effective in treating the "positive" symptoms of schizophrenia such as hallucinations and delusions. But they did not significantly affect the negative, or cognitive, symptoms such as withdrawal and apathy. Kandel, *In Search of Memory*, 357. See also Raymond J. Friedman, John G. Gunderson, and David B. Feinsilver, "The Psychotherapy of Schizophrenia: A NIMH Program" (paper presented at the annual meeting of the American Psychiatric Association, Dallas, Texas, May 1–5, 1972).

25.   Raymond J. Friedman to Alfred Goldberg, Aug. 26, 1971, author's collection.

26.   Milton Wexler to Hussain Tuma, 1971, author's collection. Working group members presented their findings at the annual meeting of the American Psychiatric Association in May 1972, a workshop at Austen Riggs in October 1972, and at the midwinter meeting of the American Psychoanalytic Association in New York a few months later, although Dad did not attend the latter.

27.   Milton Wexler to David Feinsilver, 1971, author's collection.

28.   That Jennifer Jones Simon, his patient, headed the foundation that sponsored the trip and was part of the delegation did not seem to faze my father, although it once again raised ethical questions.

29.   Milton Wexler, "On the Nature and Treatment of Schizophrenia" (unpublished manuscript, 1981), author's collection.

30.   Robert Wallerstein, interview with the author, Belvedere, California, Jun. 28, 2014. Nonetheless, some readers in the late twentieth century did take note of his papers, including a psychologist named Daniel Schiff, who wrote to Dad in 1990 saying that he had found Dad's name coming up repeatedly in the literature on the psychoanalytic treatment of schizophrenia. "Amid a lot of technically dry stuff," he wrote, "I found your papers rich soil loaded with precious insights." Daniel Schiff to Milton Wexler, Feb. 15, 1990, author's collection.

31.   See Benedict Carey, "New Approach Advised to Treat Schizophrenia," *New York Times*, Oct. 20, 2015; Benedict Carey, "New Plan to Treat Schizophrenia Is Worth Added Cost, Study Says," *New York Times*, Feb. 1, 2016; John M. Kane, Delbert G. Robinson, Nina R. Schooler, Kim T. Mueser, David L. Penn, Robert A. Rosenheck, Jean Addington, et al., "Comprehensive Versus Usual Community Care for First Episode Psychosis: Two-Year Outcomes from the NIMH RAISE Early Treatment Program," *American Journal of Psychiatry* 173, no. 4 (2016): 362–72; John G. Gunderson, Arlene F. Frank, Howard M. Katz, Marsha L. Vannicelli, James P. Frosch, and Peter H. Knapp, "Effects of Psychotherapy in Schizophrenia: II. Comparative Outcome of Two Forms of Treatment," *Schizophrenia Bulletin* 10, no. 4 (1984): 564–98.

32   Elyn R. Saks, *The Center Cannot Hold: My Journey Through Madness* (New York: Harper & Row, 2007), 331.

7. FREUDIAN FATHERS AND (PROTO-) FEMINIST DAUGHTERS

1.   Elinor Yahm, "I Look at Life from Both Sides Now." In *Growing Up Observed: Tales from Analysts' Children*," ed. Herbert S. Strean (New York: Haworth, 1987), 33.

2.   Milton Wexler, "The Structural Problem in Schizophrenia: Therapeutic Implications," *International Journal of Psychoanalysis* 32, no. 3 (1951): 164.

3. Milton Wexler to "Sundry Parents," memo, Nov. 14, 1996, author's collection; Julie V. Iovine, "When Parents Take Charge," *New York Times*, Nov. 7, 1996, https://www.nytimes.com/1996/11/07/garden/when-parents-take-charge.html.

4. Milton Wexler, "Psychotherapy of Schizophrenia," (lecture presented at LAPSI, Oct. 13–20, 1958); Milton Wexler, "Psychotherapy of Schizophrenia" (lecture presented at the University of Southern California School of Medicine, Dec. 9, 1964).

5. Rachel Devlin, *Relative Intimacy: Fathers, Adolescent Daughters, and Postwar American Culture* (Chapel Hill: University of North Carolina Press, 2005), 110–12.

6. In the unedited manuscript of *RVM*, my father described the tensions between clinical psychologists and the physicians of the American Psychoanalytic Association and the American Psychiatric Association, who asserted that psychotherapy was a medical practice and that only MDs had the authority to practice it. The state of California finally established licensure for clinical psychologists in 1967. See "Background Paper for the California Board of Psychology," Mar. 3, 2021, abp.assembly.ca.gov. See *RVM*, 480–81.

7. Committee on the Scientific and Professional Aims of Psychology, "The Scientific and Professional Aims of Psychology," *American Psychologist* 22, no. 1 (Jan. 1967): 49–76; Milton Wexler, "Psychoanalysis as the Basis for Psychotherapeutic Training," in *New Horizons for Psychotherapy: Autonomy as a Profession*, ed. Robert R. Holt (New York: International Universities Press, 1971), 184–93; Milton Wexler, "Beyond Conflict" (lecture, "Psychoanalysis and the Image of Man" conference in honor of the seventieth birthday of Theodore Reik, New York, May 13, 1958).

8. Maryline Barnard to Alice Wexler, Dec. 1, 1986, author's collection.

9. Devlin, *Relative Intimacy*, 20. The neuroscientist Eric Kandel writes, "It is difficult to capture today the fascination that psychoanalysis held for young people in the 1950s. Psychoanalysis had developed a theory of mind that gave me my first appreciation of the complexity of human behavior and of the motivations that underlie it." Eric R. Kandel, *In Search of Memory: The Emergence of a New Science of Mind* (New York: Norton, 2006), 39.

10. Sigmund Freud, "Three Essays on the Theory of Sexuality," in *Freud on Women: A Reader*, ed. Elisabeth Young-Bruehl, (New York: Norton, 1990), 136.

11. Joanne Meyerowitz, "Beyond the Feminine Mystique: A Reassessment of Postwar Mass Culture, 1946–1958," in *Not June Cleaver: Women and Gender in Postwar America, 1945–1960* (Philadelphia: Temple University Press, 1994), 229–62.

12. John S. Pearson, "Behavioral Aspects of Huntington's Chorea," in *Advances in Neurology*, vol. 1, *Huntington's Chorea: 1872–1972*, ed. André Barbeau, Thomas N. Chase, and George W. Paulson (New York: Raven, 1973), 710.

13. Milton Wexler to Henry Wexler, Jun. 5, 1957, Wexler family papers.

8. REVELATIONS

1. Milton Wexler to Alice Wexler, Jul. 12, 1971, author's collection.

2. Alice Wexler, diary entries, Jan. 1986, Jan. 5, 1991, author's collection.

3. Romi Greenson to Alice Wexler, Apr. 27, 1964, author's collection.

4. William Hackman, *Out of Sight: The Los Angeles Art Scene of the Fifties* (New York: Other, 2015), 98.

5. See, for example, Kellie Jones, *South of Pico: African American Artists in Los Angeles in the 1960s and 1970s* (Durham, NC: Duke University Press, 2017); Sarah Schrank, *Art and the City: Civic Imagination and Cultural Authority in Los Angeles* (Philadelphia: University of Pennsylvania Press, 2009); Hunter Drohojowska-Philp, *Rebels in Paradise: The Los Angeles Art Scene and the 1960s* (New York: Henry Holt, 2011); Kristine McKenna, *The Ferus Gallery: A Place to Begin* (Göttingen, Germany:

Steidl, 2009); Stephanie Barron, Sheri Bernstein, and Ilene Susan Fort, eds., *Reading California: Art, Image, and Identity, 1900–2000* (Los Angeles: University of California Press, 2000).

6.   Ed Moses, interview with the author, Apr. 18, 2014, Los Angeles, California.

7.   Milton Wexler, "John Altoon: Personal Recollections," in *John Altoon* (San Diego: Museum of Contemporary Art, 1997), 11; Roberta (Babs) Thomson, interview with the author, Oct. 9, 2015, London.

8.   Milton Wexler to Lawrence Kubie, Feb. 9, 1967; Milton Wexler and Ralph Greenson, "Creativity in the Graphic Artist: An Exploration," (unpublished manuscript, n.d.); both from Lawrence S. Kubie Papers, Name File, 1916–1973, Box 63, Library of Congress, Washington, DC; Milton Wexler, "Aggression and the Creative Impulse," (unpublished manuscript, n.d.), author's collection. An initial draft for the study of creativity was revealing in its emphasis on "murderous competition combined with profound homosexual jealousies and envy as one major source of the creative impulse." Dad was impressed with what he called the periodicity of creativity in the group of mostly young white male artists who came to see him who all emerged around the same time. In his view they shared similar love-hate relations with their peers, "suffered wild admixtures of elation and guilt at the death or defeat of a fellow artist, stole liberally from each other while attempting also to develop unique styles, upstaged each other in dress, mannerisms, or storytelling, while defending the group to the death from outside intrusions." Successful creative spurts also spread throughout the group. Nothing came of this proposal, which seems more a description of this specific cohort of artists in 1960s Los Angeles than a view into the creative process more broadly.

9.   Ed Moses, interview with the author, Apr. 18, 2014, Los Angeles, California.

10. *RVM*, 295. See also Milton Wexler to George S. Klein, Mar. 7, 1967; Milton Wexler, "Project: Creativity in the Graphic Artist: An Exploration," memorandum, Feb. 6, 1967; both from George S. Klein Papers, Drs. Nicholas and Dorothy Cummings Center for the History of Psychology, University of Akron, Akron, Ohio.

11. Frank Gehry, telephone conversation with the author, Jan. 12, 2012, Santa Monica, California.

12. Frank Gehry, telephone conversation with the author, Jan. 12, 2012, Santa Monica, California. See also Paul Goldberger, *Building Art: The Life and Work of Frank Gehry* (New York: Knopf, 2015), esp. 136–40.

13. Milton Wexler, conversation with the author, Oct. 3, 1999, Santa Monica, California.

14. Sigmund Freud, preface to *The Interpretation of Dreams*, 2nd ed. (New York: Avon, 1965), xxvi: "It was, I found, a portion of my own self-analysis, my reaction to my father's death."

15. Milton Wexler to Henry Wexler, Nov. 8, 1962, Nov. 16, 1962, Wexler family collection.

16. Milton Wexler to Henry Wexler, Nov. 16, 1962, Wexler family collection.

17. Henry Wexler to Milton Wexler, Nov. 10, 1962, author's collection.

18. Milton Wexler to Henry Wexler, Nov. 16, 1962, Wexler family collection.

19. Milton Wexler to Henry Wexler, Nov. 16, 1962, Wexler family collection. Rejecting Kierkegaard, whom he characterized as "a loveless old hunchback," Henry suggested Montaigne "or a host of others who can offer something other than despair."

20. Milton Wexler to Henry Wexler, Nov. 16, 1962, Wexler family collection; Milton Wexler to Alice Wexler, Aug. 1964, author's collection.

21. Milton Wexler to Henry Wexler, Dec. 1962, Wexler family collection.

22. Milton Wexler to Henry Wexler, Dec. 1962, Wexler family collection.

23. Milton Wexler to Henry Wexler, [Dec. 1962], Wexler family collection.

24. Henry Wexler to Milton Wexler, Jan. 15, 1963, author's collection.

25. Milton Wexler to Henry Wexler, [Dec. 1962], Wexler family collection.

26. Milton Wexler to Henry Wexler, [Dec. 1962], Wexler family collection.

27. Milton Wexler to Henry Wexler, Jul. 10, 1964, Wexler family collection.

28. Milton Wexler to Robert Holt, Dec. 27, 1963, Robert R. Holt Papers, Drs. Nicholas and Dorothy Cummings Center for the History of Psychology, University of Akron, Akron, Ohio.

29. Milton Wexler to Alice Wexler, Jun. 1963 or 1964, author's collection.

9. THE BIG FREEDOM

1. Milton Wexler to "Dearest Ali" (Alice Wexler), Oct. 31, 1964, author's collection.
2. Milton Wexler to "Dearest Ali" (Alice Wexler), Oct. 31, 1964, author's collection.
3. Milton Wexler to "Dearest Ali" (Alice Wexler), Oct. 31, 1964, author's collection.
4. Milton Wexler to "Dearest Ali" (Alice Wexler), n.d.; Milton Wexler to "Dearest Pooch-faces" (Alice Wexler), Apr. 1965; Milton Wexler to "Hi Doll" (Alice Wexler), Apr. 1965; all from author's collection.
5. Milton Wexler to "Dearest Ali" (Alice Wexler), 1969, author's collection.
6. Alice Wexler, diary entry, Dec. 16, 1990, author's collection.
7. Milton Wexler to Lawyers Guild, KPFK, Jul. 5, 2000, author's collection.
8. Milton Wexler to Alice Wexler, Feb. or Mar. 1965, author's collection.
9. Milton Wexler to "Dearest Ali" (Alice Wexler), Feb. 1965, author's collection.
10. Ed Moses, interview with the author, Apr. 18, 2014, Santa Monica, California.
11. Milton Wexler to "Merry Christmas Pooches" (Alice Wexler and Nancy Wexler), Dec. 25, 1964, author's collection.
12. Milton Wexler to "Dearest Pooch" (Nancy Wexler), Mar. 15, 1980. Nancy Wexler collection.
13. Milton Wexler to Alice Wexler and Nancy Wexler, "Attention! Your Father Speaks!" Nov. 14, 1965; Milton Wexler to "Dearest Ali" (Alice Wexler), Nov. 20, 1965; both from author's collection.
14. Milton Wexler to "Dearest Pooch" (Alice Wexler), "Friday," [1967], author's collection.
15. Milton Wexler to "Dearest Pooch" (Alice Wexler), "Friday," [1967], author's collection.
16. Milton Wexler to "Dearest Pooch" (Alice Wexler), 1967, author's collection.
17. Milton Wexler to "Dearest Pooch" (Alice Wexler), 1965; Milton Wexler to "Dearest Alinan" (Alice Wexler and Nancy Wexler), Jan. 1965; both from author's collection.
18. Milton Wexler to Alice Wexler, Feb. or Mar. 1965; Milton Wexler to "Dearest Alianova and Naniavitch" (Alice Wexler and Nancy Wexler), Oct. 14, 1967; both from author's collection.
19. Milton Wexler to "Dearest Alinan" (Alice Wexler) and Nancy Wexler, Jan. 1965; Milton Wexler to "Dearest Ali" (Alice Wexler), Apr. 4, 1969; both from author's collection.
20. Romi Greenson to Alice Wexler, Feb. 24, 1965, author's collection.
21. Milton Wexler to "Dearest Alipoo" (Alice Wexler), Feb. or Mar. 1965, author's collection.
22. Elinor Yahm, "I Look at Life from Both Sides Now," in *Growing Up Observed: Tales from Analysts' Children*, ed. Herbert S. Strean (New York: Haworth, 1987), 34.
23. Milton Wexler to "Dearest Ali" (Alice Wexler), Apr. 4, 1969, author's collection.
24. Milton Wexler to Alice Wexler, Jul. 12, 1971, author's collection.
25. Milton Wexler to "Dearest Ali" (Alice Wexler), Jan. 1971, author's collection.
26. Milton Wexler to "Merry Christmas Pooches" (Alice Wexler and Nancy Wexler), Dec. 25, 1964; Milton Wexler to "Dearest Pooch" (Alice Wexler), Jan. 1965; Milton Wexler to "Daughterkins" (Alice Wexler and Nancy Wexler), 1965; Milton Wexler to "Dearest Alinan" (Alice Wexler and Nancy Wexler), Jan. 1965; all from author's collection.
27. Milton Wexler to Robert Holt, Dec. 25, 1964, Robert R. Holt Papers, Drs. Nicholas and Dorothy Cummings Center for the History of Psychology, University of Akron, Akron, Ohio.
28. Milton Wexler to "Merry Christmas Pooches" (Alice Wexler and Nancy Wexler), Dec. 25, 1964; Milton Wexler to "Dearest Pooch" (Alice Wexler), Jan. 1965; Milton Wexler to "Daughterkins" (Alice Wexler and Nancy Wexler), 1965; Milton Wexler to "Dearest Alinan" (Alice Wexler and Nancy Wexler), Jan. 1965; all from author's collection.
29. Milton Wexler to Alice Wexler, Apr. 1964; Milton Wexler to Alice Wexler and Nancy Wexler, "Attention! Your Father Speaks!" Nov. 14, 1965; both from author's collection.
30. Milton Wexler to "Beloved Daughters" (Alice Wexler and Nancy Wexler), Dec. 5, 1966; Milton Wexler to "Beautiful Girls" (Alice Wexler and Nancy Wexler), Oct. 22, 1967; Milton Wexler to "Dearest Alianova and Naniavitch" (Alice Wexler and Nancy Wexler), Oct. 14, 1966; Milton Wexler to Alice

Wexler and Nancy Wexler, "Random jottings to whomever with love," [1967]; all from author's collection; Milton Wexler to Abraham Maslow, Nov. 15, 1966, Abraham Maslow Collection, Drs. Nicholas and Dorothy Cummings Center for the History of Psychology, University of Akron, Akron, Ohio.

31.   Milton Wexler to "Dearest Alinan" (Alice Wexler and Nancy Wexler), [Jan. 1965]; Milton Wexler to "Dearest Poochfaces" (Alice Wexler and Nancy Wexler), [Apr. 1965]; Milton Wexler to "Dearest Dolls" (Alice Wexler and Nancy Wexler), Oct. 9, 1965; Milton Wexler to Alice Wexler, "Attention! Your Father Speaks!" Nov. 14, 1965; Milton Wexler to "Beloved Daughters" (Alice Wexler and Nancy Wexler), Dec. 5, 1966; Milton Wexler to "Beautiful Girls" (Alice Wexler and Nancy Wexler), Oct. 22, 1967; Milton Wexler to Alice Wexler and Nancy Wexler, "Random jottings to whomever with love," [1967]; all from author's collection; Allen E. Bergin and Hans H. Strupp, *Changing Frontiers in the Science of Psychotherapy* (1972; New Brunswick: Aldine Transaction, 2010), 303–10.

32.   Lisa Weinstein, telephone conversation with the author, Dec. 4, 2020. I am grateful to Alice Echols for connecting me with Weinstein.

33.   Jodie Evans to Alice Wexler, Apr. 15, 2022.

34.   Polly Howard to Nancy Wexler and Alice Wexler, [2007], author's collection; Jodie Evans to Alice Wexler, Apr. 15, 2022.

35.   Alice Wexler, diary entry, Jan. 14, 1966, author's collection.

36.   *RVM*, 262–67.

37.   Milton Wexler to "Dearest Ali" (Alice Wexler), Dec. 26, [1977], author's collection.

38.   Jean-Jacques Lebel to Milton Wexler, May 20, 1966, author's collection.

39.   My appreciation to Jean-Jacques Lebel for correcting the misconception that the play had been banned in Saint-Tropez because it called for actors to urinate on stage. It was neither banned nor did it call for urination on stage. It was performed outside Saint-Tropez because of the complexities of staging it.

40.   Milton Wexler to "Dearest Pooch" (Alice Wexler), 1967, author's collection.

41.   See Leo Rangell to Milton Wexler, Jan. 13, 1965, March 24, 1965; *Bulletin*, Los Angeles Psychoanalytic Society and Institute, Jul. 23, 1964; Milton Wexler to Leo Rangell, Jan. 1, 1965, Jan. 20, 1965, excerpted in Los Angeles Psychoanalytic Society, "The Unanimous Report of the Committee to Investigate the Behavior and Functioning of the Board of Directors Under the Motion Offered by Dr. Milton Wexler and Passed by the Los Angeles Psychoanalytic Society on April 22, 1965," Jun. 17, 1965, LAPSI archives; personal interviews and phone conversations by the author, all in Los Angeles, with Gerald Aronson (by phone), Aug. 12, 2000, Nov. 7, 2000; Maimon (Mike) Leavitt, Sep. 18, 2000, Oct. 9, 2000; Samuel Sperling, Dec. 24, 2000; Michael Gales (by phone), Aug. 17, 2000; Mel Mandel, Aug. 25, 2000; Daniel Greenson (by phone), Aug. 29, 2000; Alfred Goldberg (by phone), Nov. 6, 2000, Feb. 28, 2001; Mal Hoffs (by phone), Nov. 7, 2000; Norman Atkins, Jan. 11, 2001.

42.   Los Angeles Psychoanalytic Society, "Unanimous Report," LAPSI archives.

43.   Milton Wexler to "Dearest Poochfaces" (Alice Wexler and Nancy Wexler), Apr. 1965; Milton Wexler to "Dearest Poochikins" (Alice Wexler), May 1965; both from author's collection.

44.   Los Angeles Psychoanalytic Society, "Unanimous Report," LAPSI archives.

45.   In 2000, Douglas Kirsner, an Australian academic, published *Unfree Associations: Inside Psychoanalytic Institutes* (London: Process, 2000) with a section on what he called "The Wexler Affair." Kirsner wrote, "During [Leo] Rangell's presidency [1964–65] one of Wexler's patients complained to the society that Wexler had physically attacked her. According to the complainant, she and her analyst, Wexler, got into a fight and he had slapped her. Then, as Norman Atkins recalled, [Romi] Greenson, who shared offices with Wexler, 'ran into the office when he heard the screaming and wrestled Wexler to the ground and pulled them apart' (Atkins interview 1992)." (154). Over the course of nearly a year, the patient's complaint became the centerpiece in a procedural fight among the analysts themselves, with the patient and her grievance eventually forgotten.

      As soon as I acquired the book, this passage grabbed my attention, as I vaguely recalled that Dad had broken with LAPSI in the mid-1960s. It also raised questions in my mind because the evidence Kirsner cited for the alleged "attack" and fight appeared to consist of one interview he conducted with a former board member, Norman Atkins, thirty years afterward. The book cited no documentation from the time of the events such as correspondence, minutes, memos, or reports, from the patient nor anyone else. All citations describing the episode were for interviews in the 1990s with former LAPSI board members. When I asked Dad about these events, he acknowledged that a patient had made a complaint but denied that he had "attacked" her as Kirsner claimed and gave quite a different account of the entire episode.

Upon reading and hearing all this, the investigative journalist in me went into high gear. Over the course of nine months in 2000 to 2001, I interviewed some of the LAPSI members cited in the book, and several more. Thanks to then LAPSI president Michael Gales and acting LAPSI archivist Mal Hoffs, I was also able to access the LAPSI archives, where I found reports and correspondence relating to the events of 1964–65, material that Kirsner had not consulted and which did not support his allegations of physical abuse. Norman Atkins disavowed the statements attributed to him by Kirsner in the book, claiming he had been misrepresented. (Norman Atkins to Milton Wexler, Jan. 29, 2001). Other LAPSI board members at the time also affirmed, orally and in writing, that they had never personally heard the patient's grievance nor seen any documentation nor evidence of her complaint nor of certain other claims made by Kirsner, some of which were easily disproved. All parties agreed that no charges were ever brought against my father. Gales affirmed that "we could find no documentation of any complaint against Dr. Wexler from the patient, nor could we find any minutes of the Board acting as a grievance committee pertaining to any patient complaint." (Michael Gales to Milton Wexler, Mar. 6, 2001); see also Alice Wexler, "The 'Wexler Affair' and Historical Evidence: A Reply to Douglas Kirsner's *Unfree Associations*," *Free Associator* 5, no. 2 (Apr. 2001).

Still, no one denied that this patient had made a complaint to Leo Rangell and possibly others. But by 2000–2001, no one was clear on what it was, including my father, who had never received any written statement of it. Two letters this patient wrote to Greenson, her prior therapist, and recently archived in the Ralph R. Greenson Papers at Special Collections in the Charles E. Young Research Library at UCLA cast new light on the episode. In a 1966 letter she compared my father's treatment of her to his treatment of the patient he wrote about in a 1953 paper, "Psychological Distance as a Factor in the Treatment of a Schizophrenic Patient." He had treated that patient, she wrote, "just as he treated me. . . . The emotional atmosphere he describes [in the paper] is exactly that of my treatment." (Patient to Ralph Greenson, Jul. 25, 1966, Ralph R. Greenson Papers, Special Collections, Charles E. Young Research Library, UCLA.) Suffice it to say that the paper does not describe any physical abuse. Whether it describes verbal or emotional abuse by the standards of the time is a question I cannot answer. It does suggest trespassing "in a rude and determined fashion over the hostile barriers" of her psychotic withdrawal and ending her treatment when she did not respond, unlike her predecessor: behaviors that may have instigated the complaint.

Regarding the LAPSI board's utter failure to address the patient's grievance, whatever it may have been, she was prescient. "My grievance will still be there when this fight is over," she wrote. (Patient to Ralph Greenson, May 15, 1965, Ralph R. Greenson Papers, Special Collections, Charles E. Young Research Library, UCLA); see also Leo Rangell to Milton Wexler, Jan. 13, 1965; "The Unanimous Report," Jun. 17, 1965; *Bulletins* "From the Office of the President," Jul. 23, 1964 to Jun. 17, 1965; List of Board of Directors meetings, Jul. 5, 1964 to Jun. 8, 1965; Maimon Leavitt, Mel Mandel, and Gerald Aronson to Milton Wexler, Dec. 21, 2000; Norman Atkins to Milton Wexler, Jan. 29, 2001; Michael Gales to Milton Wexler, Mar. 6, 2001, all from LAPSI archives.

46. Milton Wexler to "Dearest Poochikins" (Alice Wexler and Nancy Wexler), [June 1965], author's collection.

47. Milton Wexler to Henry Wexler, Mar. 5, 1967, Wexler family papers.

48. Milton Wexler to Alice Wexler and Nancy Wexler, "Random jottings to whomever with love," 1967, author's collection; Milton Wexler to Henry Wexler, Mar. 5, 1967, Wexler family collection.

49. Alice Wexler, diary entry, Jan. 28, 1968, author's collection.

50. Milton Wexler to Alice Wexler and Nancy Wexler, "Memo: To Pooches Two," Jul. 20, 1968, author's collection.

10. (A) CHALLENGING FATE

1. Milton Wexler to Henry Wexler, May 1968, Wexler family collection.

2. Charles Markham to Ralph Greenson, May 24, 1968, author's collection.

3. Ntinos Myrianthopoulos, "Huntington's Chorea: The Genetic Problem Five Years Later," in *Advances in Neurology*, vol. 1: *Huntington's Chorea: 1872–1972*, ed. André Barbeau, Thomas N. Chase, George W. Paulson (New York: Raven, 1973), 155.

4. Milton Wexler to Henry Wexler, [Jul. 7, 1968], Wexler family collection.

5. *Report: Congressional Commission for the Control of Huntington's Disease and Its Consequences, Public Testimony* 4, no. 5 (1977): 71. On Woody Guthrie's last years, see Philip Buehler in collaboration with Nora Guthrie and the Woody Guthrie Archives, *Woody Guthrie's Wardy Forty: Greystone Park State Hospital Revisited* (Mt. Kisco, NY: Woody Guthrie Publications, 2013); see also Joe Klein, *Woody Guthrie: A Life* (New York: Ballantine, 1980).

6. Milton Wexler to Henry Wexler, Jun. 2, 1968, Wexler family collection.

7. Milton Wexler, "Mental Hygiene and the Muscular Dystrophy Patient," in *Proceedings of the First and Second Medical Conferences of the Muscular Dystrophy Associations of America Held in New York, NY, April 14–15, 1951 and May 17–18, 1952*, ed. Ade T. Milhorat (New York: Muscular Dystrophy Associations of America), 58–67.

8. Henry Wexler to Milton Wexler, May 24, 1968, author's collection.

9. Henry Wexler to Milton Wexler, May 18, 1968, author's collection.

10. Henry Wexler to Milton Wexler, Jul. 1968, author's collection. Watson was awarded the prize in 1962 jointly with Francis Crick and Maurice Wilkins for the discovery of the double-helix structure of DNA; Linus Pauling won in 1954 for elucidating the nature of the chemical bond; in 1937 the prize went to Albert Szent-Györgyi for discoveries related to biological combustion processes.

11. Henry Wexler, journal entry, May 18, 1968, Wexler family collection.

12. Alice Wexler, *Mapping Fate: A Memoir of Family, Risk, and Genetic Research* (Berkeley: University of California Press, 1995), 74–77.

13. Henry Wexler, journal entry, May 25, 1968, Wexler family collection.

14. Milton Wexler to Henry Wexler, Jun. 2, 1968, Wexler family collection.

15. Henry Wexler, journal entry, Jun. 20, 1968, Wexler family collection.

16. Milton Wexler to Henry Wexler, Jun. 2, 1968, Wexler family collection.

17. Milton Wexler to "Dearest Pooches" (Alice Wexler and Nancy Wexler), Sep. 7, 1968, author's collection.

18. Milton Wexler to Alice Wexler, May 27, 1969, author's collection.

19. Milton Wexler, "Autobiography" (unpublished manuscript, 2005), author's collection.

20. Milton Wexler to "Dearest Ali" (Alice Wexler), 1969, author's collection.

21. Milton Wexler to Henry Wexler, 1969, Wexler family collection.

22. Dad was able to get free legal services from a Los Angeles law firm, Alef and Schnitzer.

23. Milton Wexler to Charles Markham, Jan. 6, 1969, author's collection.

24. "I think it was largely a calculated, Pied Piper approach," Roberts told me, "strictly Skinnerian." He and Dad talked about it in those terms. As Roberts put it, Dad provided an attractive social setting, an attractive peer group (including himself), a chance to mingle with glamorous people, and an opportunity to come to Southern California for a weekend. Roberts in turn introduced Dad to a creative young psychologist at Harvard named Steven Matthysse, who proposed that Huntington's

was the motor expression of the same process that causes mental changes in schizophrenia since both involve uncontrolled movement: one in the body, the other in the mind. He thought Huntington's showed how certain diseases at certain times become teachers, illuminating function beyond their clinical significance. Matthysse also joined the science advisory board (Eugene Roberts, interview with the author, Dec. 1990, Duarte, California) Steve Matthysse, interview with R. Steve Uzzell, May 1979.

25. Milton Wexler to Lewis Levin, Mar. 16 and 28, 1970; Charles Kleeman and Milton Wexler to William J. Dreyer, Apr. 2, 1970; both from HDF archive. Early scientific advisory board members included neurologists Charles Markham and Edward Davis, the latter being the neurologist who diagnosed Mom. There were also clinicians who had no expertise with Huntington's but were willing to help, such as John Menkes, a pediatric neurologist at UCLA; also two clinicians at Cedars-Sinai whom Dad believed could give the proposed Huntington's center leverage within the institution: Charles Kleeman, an internist and the chief of medicine, and Ronald Okun, a pharmacologist. Dad's psychoanalyst friends Romi Greenson and the ex-Topekan Gerry Aronson became members along with Bernard Salick, an entrepreneurial young physician who was a medical school friend of Romi's son, Daniel. And through Markham, Dad recruited established Huntington's investigators John Whittier, Ntinos Myrianthopoulos, and André Barbeau. From his discussions with both HD family members and researchers, he learned—and would later emphasize—that several other projects were crucially needed, such as an information exchange to make accurate knowledge about the diagnosis and treatment of Huntington's available to medical practitioners who were often woefully ill informed. Additionally he proposed to create materials for affected families to acquaint them with what few resources were available (Milton Wexler to Lewis Levin, Mar. 16, 1970, HDF archive; Milton Wexler to Mr. Trainor, May 14, 1970, author's collection). One thread that ran through his letters at this time and became the hallmark of my father's thinking about science advocacy was his support for "the free exchange of knowledge, hunches, [and] hypotheses" that he believed would lead to "the secret hiding place of the answers" (Milton Wexler to John S. O'Brien, Feb. 3, 1970, author's collection). Along with assembling a scientific advisory board, Dad wanted—or needed—to establish a board of trustees to guide the fundraising and general operations of the chapter, a task that did not really get under way until after Dad had flown to New York to meet with Marjorie Guthrie and her associates at CCHD.

26. Henry Wexler, journal entry, Dec. 11, 1969, Wexler family collection.

27. Milton Wexler to Lewis Levin, Mar. 16, 1970, HDF archive. CCHD's New York chapter's arts council counted thirty-three members as of April 1, 1971, including such distinguished artists as Joan Baez, Mel Brooks, Judy Collins, Elaine May, Pete Seeger, and Stephen Sondheim, to name just a few. Marjorie Guthrie to Milton Wexler, Apr. 1, 1971, HDF archive.

28. The five members in 1970 were Maryline Barnard, Marjorie Fasman, Sandy Ruben, Bernard Salick, and Milton Wexler (California CCHD chapter, minutes of general membership meetings, Jan. 30, 1970, n.d., Mar. 13, 1972), HDF archive. New members were Elaine Attias, Mrs. Morton Brandler, Marvin Garfield, Frank Gehry, and Marjorie Guthrie. Minutes of meetings during this time did not always name those present, but the California CCHD chapter's stationery listed thirty-six members by March 13, 1972.

29. California CCHD chapter, minutes of general membership meeting, Apr. 20, 1971, HDF archive. See Stephen Farber and Marc Green, *Hollywood on the Couch: A Candid Look at the Overheated Love Affair Between Psychiatrists and Moviemakers* (New York: William Morrow, 1993), 215–39.

30. In fact, Altoon had already given Nancy and me each a drawing.

31. Milton Wexler to "Dearest Dolls" (Alice Wexler and Nancy Wexler), Feb. 13, 1969, author's collection.

32. Dad acquired eight drawings by Altoon over the years and loaned several of them to exhibitions of the artist's work, including one in 1971 at the Whitney Museum of American Art in New York, curated by Walter Hopps, and another at the San Diego Museum of Art from Dec. 7, 1997 to Mar.

11, 1998, for which Dad wrote a brief remembrance included in the catalog. Many artists and curators who wrote about Altoon's work credited Dad with helping to stabilize his mental state and enable him to work. Walter Hopps and Elke Solomon, *John Altoon: Drawings and Prints* (New York: Whitney Museum of American Art, 1971); David Stuart, "John Altoon," catalog for the March 1980 de Saisset Museum exhibition (Santa Clara, California: University of Santa Clara, 1980), 16; Carol S. Eliel, "John Altoon: Confronting the Void," in *John Altoon*, ed. Carol Eliel (Los Angeles: Los Angeles County Museum of Art, 2014).

33.   Milton Wexler to "Dearest Dolls" (Alice Wexler and Nancy Wexler), Feb. 13, 1969, author's collection; Milton Wexler to Gerald Nordland, 1967, quoted in Stuart, "John Altoon," 13.

11. WORKSHOPS OF THE POSSIBLE

1.   François Jacob, *The Statue Within: An Autobiography* (Cold Spring Harbor, NY: Cold Spring Harbor Press, 1995), 296–97.

2.   Tickets at the Bowl usually ranged from $5 to $20. However, the California CCHD chapter reserved forty boxes to sell to "patrons" at $1,000 per ticket, with a bank of other seats selling for $15, $25, $50, and $100, for a total of two thousand seats sold at "premium" prices. Dad aimed to earn $150,000 from the event. A company called Sight and Sound Productions offered to help stage the event without charge, and the California CCHD chapter agreed to organize advertising, secure clearances, and sell tickets. Milton Wexler to Arthur Alef, 1970, HDF archive.

3.   The situation regarding rights was confusing. Three days before the concert, Frank Wells, the vice-president of Warner Brothers at the time, signed an agreement with Dad to the effect that although not all rights had been secured or terms agreed on for a record or film, "neither of us shall have the right to distribute or otherwise exploit either such film or such sound track until and unless a satisfactory agreement is entered into between us after such concert" (Frank G. Wells to Committee to Combat Huntington's Disease, Sep. 9, 1970, HDF archive). However, according to Harold Leventhal, rights "for film and television usage of the songs are controlled by United Artists Inc." so that the taping at the Hollywood Bowl was in effect "speculative" since "none of us can guarantee all clearances." At the same time Leventhal affirmed that "all rights to this taped film [are] held by the Committee to Combat Huntington's Disease, Inc., subject to final approval for its usage" (Harold Leventhal to John Calley, Aug. 12, 1970, HDF archive). Two weeks later he affirmed to a Warner Brothers official that the "tape recordings of the concert shall be the property of the Committee to Combat Huntington's Disease, Inc." (Harold Leventhal to Joseph Smith, Aug. 27, 1970, HDF archive).

4.   "We're off to the races in this thing," Dad wrote to me early in the summer of 1970. "My head spins with all this production stuff but I'm gradually learning the ropes." Milton Wexler to Alice Wexler, [Jul. 1970], author's collection.

5.   Frank Gehry, conversation with the author, Sep. 4, 2021. Gehry had recently designed the installation of Sonotubes, enormous cardboard tubes, for the stage as a temporary move to improve acoustics at the Bowl.

6.   Marjorie Guthrie to Milton Wexler, Sept. 29, 1970, HDF archive.

7.   There is considerable correspondence relating to this controversy. See for example Harold Leventhal to Joseph Smith, Aug. 27, 1970; Frank G. Wells to Committee to Combat Huntington's Disease, Inc., Sep. 9, 1970; Marjorie Guthrie to Milton Wexler, Oct. 5, 1971; Milton Wexler, "Memorandum of conversations concerning the Hollywood Bowl Record," Jan. 20, 1972; Milton Wexler to Jay Tolson, Oct. 7, 1972, all from the HDF archive.

8.   Milton Wexler to Jay Tolson, Oct. 7, 1972, HDF archive.

9.   Milton Wexler to "Dolls" (Alice Wexler and Nancy Wexler), Oct. 27, 1972, author's collection.

10. Board of Trustees, meeting minutes, Feb. 28, 1973, HDF archive. The agreement between CCHD and the California chapter is dated July 15, 1974.

11. HDF scientific advisory board, meeting minutes, Jun. 21, 1972, HDF archive. The minutes indicate that the idea for a research center was that of John Menkes.

12. Eugene Roberts, interview with the author, Dec. 1990, Los Angeles, California. Roberts said, "I don't think there's anything like an expert—there are good scientists and bad scientists. If a good scientist undertakes to look at a problem, he'll look at it as it should be looked at.... And to bring in an old-line expert who's worked with this thing for twenty years, who has written a book on it, is nowheres-ville as far as I'm concerned."

13. Milton Wexler to "Dearest Pooches" (Alice Wexler), Mar. 1971, author's collection; Milton Wexler, report, Apr. 4, 1971, HDF archive.

14. Elie Shneour to Milton Wexler, Mar. 13, 1971, HDF archive.

15. Milton Wexler to Marjorie Guthrie, Dec. 6, 1971, HDF archive.

16. Milton Wexler to John Whittier, Dec. 15, 1970, HDF archive.

17. Milton Wexler to "Dearest Pooches" (Alice Wexler and Nancy Wexler), Mar. 1971, author's collection.

18. Milton Wexler to Alice Wexler, Nov. 26, 1983, author's collection.

19. Milton Wexler, conversation with the author, Santa Monica, California, 1984.

20. Hans-Jörg Rheinberger and Staffan Müller-Wille, trans. Adam Bostanci, *The Gene: From Genetics to Postgenomics* (Chicago: University of Chicago Press, 2017), 79; HDF scientific advisory board, meeting minutes, Feb. 28, 1973, HDF archive; HDF executive committee, meeting minutes, Jun. 17, 1986, HDF archive.

21. "Summary of a Workshop," Jan. 21–22, 1989, Frederick R. Weisman Art Foundation.

22. Milton Wexler to "Dearest Ali" (Alice Wexler), Sep. 30, 1971, Jun. 28, 1971, author's collection.

23. Milton Wexler to "Dearest Dolls" (Alice Wexler and Nancy Wexler), 1972, author's collection.

24. Milton Wexler to "Doll" (Alice Wexler), n.d., author's collection.

25. Milton Wexler to "Dearest Dolls" (Alice Wexler and Nancy Wexler), Sep. 28, 1969, author's collection.

26. Milton Wexler to "Dearest Dolls" (Alice Wexler and Nancy Wexler), 1969 or 1970, author's collection; HDF scientific advisory board, meeting minutes, Feb. 28, 1973, HDF archive.

27. Milton Wexler to "Dearest Dolls" (Alice Wexler and Nancy Wexler), 1969 or 1970, author's collection.

28. Milton Wexler to Henry Wexler, Aug. 1973, author's collection.

29. Milton Wexler to "Dearest Ali" (Alice Wexler), 1969, author's collection.

30. Milton Wexler to Alice Wexler, Jan. 1970, Jan. 16, 1970, author's collection.

31. Milton Wexler to "Dearest Pooch" (Nancy Wexler), Oct. 18, 1971, author's collection.

32. Ilse trained and exhibited dogs, went to the opera and theater, traveled to Europe and Canada, and had many friends. What made such a life possible? Beyond the devotion and considerable resources of her parents, Dad explained that Ilse was treated "with utmost respect" as an adult early on, that she was made an active participant in every aspect of her care, and that "although a considerable indulgence was present, so also was there a considerable strictness in certain areas—about health reg-imens, hours of sleep, etc. A no-nonsense attitude." She was given many responsibilities and taught about her treatment. Whatever pity people may have felt for her was "never evident in a single ges-ture or tone." Milton Wexler, lecture given to the Muscular Dystrophy Association, Santa Monica, California, Oct. 27, 1971.

33. Milton Wexler to Fay Walters, Mar. 4, 1996, author's collection.

34. Lillian Hellman to Milton Wexler, Mar. 27, 1978, author's collection.

35. Lillian Hellman to Milton Wexler, Jun. 12, 1978, author's collection.

36. *RVM*, 329.

37. Peter Feibleman, *Lilly: Reminiscences of Lillian Hellman* (New York: William Morrow, 1988), 218.

38. Lillian Hellman to Milton Wexler, May 12, 1980, author's collection.

39. Dad claimed that Glen Close's character in the film *Fatal Attraction* was a realistic portrayal of a certain kind of empty, voracious, destructive woman rather than a misogynistic stereotype, precipitating a heated argument between us. Alice Wexler, diary entry, Jun. 30, 1991, author's collection.

40. Lillian Hellman to Milton Wexler, Oct. 1, 1980, author's collection.

41. Lillian Hellman to Alice Wexler, Sep. 26, 1980, author's collection.

42. Lillian Hellman to Milton Wexler, Oct 28, 1980, author's collection.

43. Milton Wexler, "Reminiscence" (unpublished manuscript, n.d.), author's collection.

44. Milton Wexler to Lillian Hellman, Jun. 6, 1983, author's collection.

45. Milton Wexler to Eppie Lederer, Oct. 15, 1996, Nov. 14, 1996, author's collection. See also *RVM*, 328–30.

46. *RVM*, 329; Milton Wexler to Eppie Lederer, Oct. 15, 1996, author's collection.

12. MAKING FRIENDS, MAKING LOVE

1. James F. Gusella, Nancy S. Wexler, P. Michael Conneally, Susan L. Naylor, Mary Anne Anderson, Rudolph E. Tanzi, Paul C. Watkins, et al., "A Polymorphic DNA Marker Genetically Linked to Huntington's Disease," *Nature* 306, no. 5940 (Nov. 17, 1983): 234–38.

2. See Alice Wexler, *Mapping Fate: A Memoir of Family, Risk, and Genetic Research* (Berkeley, CA: University of California Press, 1995).

3. François Jacob, *The Statue Within: An Autobiography* (Cold Spring Harbor, NY: Cold Spring Harbor Press, 1995), 16.

4. Gillian P. Bates, "The Molecular Genetics of Huntington Disease—a History," *Nature Reviews Genetics* 6 (2005): 767.

5. HDF scientific advisory board, meeting minutes, Jan. 9, 1984, HDF archive.

6. Anne Young to Milton Wexler, Jan. 18, 1991, Nancy Wexler collection.

7. Alice Wexler, diary entry, Dec. 2, 1996, author's collection.

8. Because the abnormal (expanded) huntingtin gene has close to 100 percent penetrance—meaning that if you have the expanded version, you will at some point develop symptoms if you live long enough—the presymptomatic genetic test differs from most such tests, which typically indicate an increased or lower-than-average risk for a specified condition or disease.

9. Milton Wexler to Alice Wexler and Nancy Wexler, Nov. 15, 1971, author's collection.

10. Milton Wexler to Alice Wexler and Nancy Wexler, Nov. 27, 1977, Feb. 28, 1983, author's collection.

11. Eppie Lederer, "Huntington's Disease: Here Are the Facts," *Chicago Sun-Times*, Oct. 10, 1979.

12. Carol Felsenthal, "Dear Ann," *Chicago Magazine*, Feb. 1, 2003, 17.

13. Eppie Lederer to Milton Wexler, Mar. 7, 1988, author's collection.

14. Eppie Lederer to Milton Wexler, Apr. or May 1981, author's collection.

15. Alice Wexler, diary entry, 1977, author's collection.

16. Eppie Lederer to Milton Wexler, May 3, 1981, author's collection.

17. Milton Wexler to Eppie Lederer, n.d., author's collection.

18. Milton Wexler to Eppie Lederer, 1989, author's collection; Milton Wexler to Eppie Lederer, Sep. 15, 1990, author's collection.

19. Milton Wexler to Eppie Lederer, Nov. 13, 1990, author's collection.

20. Milton Wexler to Eppie Lederer, [1996], author's collection.

21. *Storm of Strangers*, 1969, directed by Ben Maddow; *The Dreamer that Remains: A Portrait of Harry Partch*, 1972, directed by Stephen Pouliot; *Italianamerican*, 1974, directed by Martin Scorsese.

22. Elaine Attias to Nancy Wexler, Apr. 7, 1975, Nancy Wexler collection.

23. Alice Wexler, diary entry, Jun. 5, 1975, author's collection.

24. Milton Wexler to Elaine Attias, Oct. 1982, courtesy of Dan Attias.

25. Milton Wexler to Elaine Attias, 1982, Jun. 1985, courtesy of Dan Attias.

26. Milton Wexler to Elaine Attias, May 11, 1985, courtesy of Dan Attias.

27. Milton Wexler to Elaine Attias, Apr. 23, 1986, courtesy of Dan Attias.

28. Milton Wexler to Elaine Attias, Sep. 14, 1990, courtesy of Dan Attias.

29. "Libby" (unpublished manuscript, n.d.), 17–18, author's collection.

30. *RVM*, 164.

31. Henry Wexler to Milton Wexler, Mar. 23, 1969, Wexler family collection. Henry was reminded of Bernard Malamud's *A New Life*. "It almost proves your point," he wrote to Dad. "I say almost because when you close the book you feel that the miserableness of intimacy is still better than . . ." And there the letter becomes illegible.

32. Milton Wexler to Nancy Wexler, n.d., Nancy Wexler collection.

33. Milton Wexler to Alice Wexler and Nancy Wexler, Feb. 28, 1983, author's collection; Milton Wexler to Richard Atlas, Aug. 7, 2000, author's collection.

13. RETELLING LIVES

1. Julie Andrews with Emma Walton Hamilton, *Home Work: A Memoir of My Hollywood Years* (New York: Hachette, 2019). According to Dad, the scene never made it into the film because it was too naturalistic and clashed in tone with the rest of the film.

2. Peter Rainer, " 'Man Who Loved Women' Overloaded with Affection," *Los Angeles Herald Examiner*, Dec. 16, 1983; Charles Champlin, "Edwards, Hollywood in Detente," *Los Angeles Times*, Dec. 1, 1983; Vincent Canby, "A Bountiful Fall Season for New York Filmgoers; 'That's Life!' by Blake Edwards," *New York Times*, Sep. 26, 1986; Victor Zak, "Crisis on the Home Front," *Dispatch* (Hudson/Bergen Counties, NJ), Sep. 29, 1986.

3. Stephen Farber and Marc Green, *Hollywood on the Couch: A Candid Look at the Overheated Love Affair Between Psychiatrists and Moviemakers* (New York: William Morrow, 1993), 236–37; Alice Wexler, diary entry, Aug. 20, 1998, author's collection.

4. He rarely dated his drafts, which typically went through many revisions, so the chronology of his scripts and treatments remains unclear. Two brief early treatments were "Post-mortem Blues," about regrets after death, and "Bad Apple," about the consequences of poor parenting.

5. Farber and Green, *Hollywood on the Couch*, 232.

6. Dad described "Penelope" as a fairy tale about overcoming: the victim becoming the victor.

7. Milton Wexler to "Dearest Ali" (Alice Wexler), Oct. 31, 1964, author's collection.

8. Milton Wexler to Lawrence Taubman, Jun. 12, 1995, author's collection.

9. *RVM*, 167–185.

10. Milton Wexler to Alice Wexler, Jun. 1963, author's collection.

11. Milton Wexler, "Memo to: Dewanna, Ruth, and Arthur," Jul. 5, 1990, author's collection; Alice Wexler, diary entry, Jun. 10, 1979, author's collection.

12. Alice Wexler, diary entry, Aug. 20, 1998, author's collection; Milton Wexler to Nancy Wexler, Mar. 15, 1980, Nancy Wexler collection.

13. Alice Wexler, diary entry, Mar. 2, 2006, author's collection.

14. Milton Wexler to "Dearest Poochfaces" (Alice Wexler and Nancy Wexler), Apr. 1965; Mardi Horowitz to Milton Wexler, Nov. 11, 1985; ACJB to Alice Wexler, Apr. 16, 2007; all from author's collection.

15. By *art*, Duchamp was not referring to physical beauty, according to Dad. Rather he was referring to "the shaping of his mind, of his character, and of his life into a pattern he felt to be joyous and free." "WW," 27.

16. Milton Wexler to "Dearest Ali" (Alice Wexler), Jan. 1971, author's collection.

17. "WW," 52.

18. "WW," 38, 87; Micki McGee, *Self-Help, Inc.: Makeover Culture in American Life* (Oxford: Oxford University Press, 2005), 22–23.

19. "WW," 131.

20. Milton Wexler to Elaine Attias, Sep. 12, 1989, courtesy of Dan Attias.

21. Milton Wexler to Alice Wexler, Sep. 9, 1986, author's collection.

22. Alice Wexler, diary entry, Dec. 16, 1990, author's collection.

23. Milton Wexler to Alice Wexler, Jan. 21, 1991, author's collection.

24. For a discussion of these issues, see, for instance, Paul John Eakin, ed., *The Ethics of Life Writing* (Ithaca, NY: Cornell University Press, 2004).

14. LIFE UNDERWATER

1. Alice Wexler, diary entry, Feb. 23, 1975, author's collection.

2. Milton Wexler to Alice Wexler, Mar. 21, 1988, author's collection.

3. Marylouise Oates, "Star Analyst Gets the Star Treatment," *Los Angeles Times*, Sep. 26, 1988.

4. HDF board of trustees, meeting minutes, Oct. 4, 1989, HDF archive; David Callahan, *The Givers: Wealth, Power, and Philanthropy in a New Gilded Age* (New York: Knopf, 2017).

5. Milton Wexler to Elaine Attias, Nov. 15, 1989, courtesy of Dan Attias.

6. Milton Wexler to Elaine Attias, Sep. 12, 1989, courtesy of Dan Attias.

7. Milton Wexler to Nancy Wexler, Apr. 28, 1988, Nancy Wexler collection.

8. Alice Wexler, diary entry, Mar. 15, 1990, author's collection.

9. Alice Wexler, diary entry, Jul. 26, 1992, author's collection.

10. Sigmund Freud, quoted in Peter Gay, *Freud: A Life for Our Time* (New York: Norton, 1988), 610.

11. Milton Wexler, "Notes on Psychotherapy in Schizophrenia" (lecture presented at a meeting of the Southern California branch of the American Psychiatric Association, Los Angeles, Feb. 6, 1953).

12. Alice Wexler, diary entry, May 19, 1991, author's collection.

13. Elisabeth Young-Bruehl, *Anna Freud: A Biography* (New York: Norton, 1988), 138–39.

14. Natalie Angier, "Ten-Year Search Leads Medicine to Elusive Gene," *New York Times*, Mar. 24, 1993; Huntington's Disease Collaborative Research Group, "A Novel Gene Containing a Trinucleotide Repeat that Is Expanded and Unstable on Huntington's Disease Chromosomes," *Cell* 72, no. 6 (Mar. 26, 1993): 971–83.

15. For books highlighting the gene search story and Nancy's role in it, see Jerry E. Bishop and Michael Waldholz, *Genome: The Story of the Most Astonishing Scientific Adventure of Our Time— the Attempt to Map All the Genes in the Human Body* (New York: Simon and Schuster, 1990); Jeff Lyon and Peter Gorner, *Altered Fates: Gene Therapy and the Retooling of Human Life* (New York: Norton, 1996); see also Joseph B. Martin, *Alfalfa to Ivy: Memoir of a Harvard Medical School Dean* (Edmonton: University of Alberta Press, 2011), 155–86; Robert Cook-Deegan, *The Gene Wars: Science, Politics, and the Human Genome* (New York: Norton, 1994), 231–55.

16. Alice Wexler, diary entry, Oct. 17, 1996, author's collection.

17. Alice Wexler, diary entry, Nov. 1997, author's collection.

15. THE OLD LEAF

1. Milton Wexler to Alice Wexler and Nancy Wexler, Nov. 7, 1994; Milton Wexler, "Psychological Insurance Plan, GAPING" (unpublished manuscript, n.d.); both from author's collection.

2. Alice Wexler, diary entries, Feb. 27, 2007; Mar. 1, 2007; Mar. 5, 2007; Mar. 11, 2007, author's collection.

3. Elaine Woo, "A Visionary Who Led a Genetic Revolution," *Los Angeles Times*, Mar. 22, 2007; Douglas Martin, "Milton Wexler, Groundbreaker on Huntington's, Dies at 98," *New York Times*, Mar. 24, 2007.

4. Jean-Jacques Lebel, interview with the author, Oct. 2, 2015, Paris.

5. Frank Gehry, conversation with the author, Jan. 12, 2012, Los Angeles.

EPILOGUE

1. Milton Wexler to Alice Wexler and Nancy Wexler, Nov. 7, 1994, author's collection.

Index

insulin shock therapy, 7
interdisciplinary workshops, 164–67, 267n12
International Design Conference (Aspen), 168
International Journal of Psychoanalysis, 13
International Psychoanalytical Association,
    13; congress of, in Amsterdam, 137; in
    Copenhagen, 139–40; in Rome, 93
Interpretation of Dreams, The (Freud), 84

Jacob, François, 160, 179
Jet Propulsion Laboratory, 167
Jews, in Brooklyn, 18–19, 249n18; in Los Angeles,
    80; at Menninger, 56, 80
Jones Simon, Jennifer, 77, 95, 167, 258n28
Journal of Heredity, 35

Kandel, Eric, 94, 257n22, 259n9
Kerouac, Jack, 115, 117
Khan, Masud, 138
Kienholz, Edward, 108
Kirsner, Douglas, 262n45
Kleeman, Charles, 265n25
Klopfer, Bruno, 40
Knight, Robert P., 6–7, 55, 72, 245n3, 246n21
Korean War, 73
Krout-Hasegawa, Ellen, 199

La catira (the blonde), 179
La Cienega Boulevard, Los Angeles, 107–8, 110
Lake Maracaibo, 178–79, 218
Lake Tahoe, 87–89, *88*
Lampell, Millard, 161
Landers, Ann. *See* Lederer, Eppie
LAPSI. *See* Los Angeles Psychoanalytic Institute
law, Milton Wexler, study of, 28–32
Law for the Prevention of Hereditarily Diseased
    Progeny, 34
Law Review (journal), 32
lay analysts, 37–38, 47, 73, 77, 90, 252n33; Freud
    and, 38; UCLA and, 90
Lazar, Irving "Swifty", 19
Leavitt, Mike, 57
Lebel, Jean-Jacques, 138, 139, 140, 225–26
Lederer, Eppie, 185–89, *187*
Leiris, Michel, 139–40
Leites, Nathan, 134
Lenin, Vladimir, 96
Leventhal, Harold, 160, 163
"Libby" (Milton Wexler), 192–93

libido, 15–16, 47, 62, 253n15
Lightning Field, The (land art), 210
Lilly (Feibleman), 173
Lindner, Robert, 48, 123
Listening with the Third Ear (Reik), 43
lobotomy, 7
Look Through the Rearview Mirror, A (Milton
    Wexler), 47, 197–98, 199
Los Angeles, 75, 76, 79–83, 107–10; art
    scene of, 107–10, 153; art schools of, 108;
    McCarthyite atmosphere of 1950s, 76;
    psychoanalysis in, 77
Los Angeles Psychoanalytic Institute (LAPSI),
    77, 91, 92–93, 100–1, 117; board of directors
    of, 141–42; Milton Wexler rift with, 140–41,
    262–63n41–46
Lowry, Bates, 138–39
Lundberg, Ferdinand, 102

macular degeneration, 213
Magister Ludi (Hesse), 134
Man Who Loved Women, The (film), 195–96
Mapping Fate (Alice Wexler), 2, 150, 203, 206
Marine Biological Laboratory (Woods Hole), 34
Markham, Charles, 145, 154, 165, 265n25
marriage, Huntington's disease and prohibitions
    against, 34–35; Leonore Sabin's delays of,
    35–36; views on: of Alice, 104, of Henry
    Wexler, 36, 150; of Milton Wexler,106–7, 118,
    153, 164; of Nathan Wexler, 22–23
Marshall, Arthur, 55–56
Martha Graham Dance Company, 147
Martha's Vineyard, 173, 174
Maslow, Abraham,134, 247n5, 251n19
Masochism in Modern Man (Reik), 45
Massachusetts General Hospital, 181
May, Elaine, 113, 136
May, Philip R. A., 90–91, 256n9, 258n26;
    schizophrenia research of, 90–92
Mayman, Martin, *72*
Mazziotta, John, 209
McCarthyism, 76
McKenna, Anne, 210
Mead, Margaret, 79
Mead, Taylor, 139
medical profession, views on Huntington's
    disease of, 34–36, 150; tensions with clinical
    psychologists, 100, 259
Memoirs of My Nervous Illness (Schreber), 15